THE TWO-WHEEL TRACTOR HANDBOOK

SMALL-SCALE EQUIPMENT AND INNOVATIVE TECHNIQUES FOR BOOSTING PRODUCTIVITY

ZACH LOEKS

Cover design by Diane McIntosh.

Cover photos by Jedediah Loeks

Printed in Canada. First printing December 2022.

Any other inquiries can be directed by mail to:
New Society Publishers
P.O. Box 189, Gabriola Island, BC V0R 1X0, Canada
(250) 247-9737

LIBRARY AND ARCHIVES CANADA CATALOGUING IN PUBLICATION

Title: The two-wheel tractor handbook : small-scale equipment and innovative techniques for boosting productivity / Zach Loeks.
Names: Loeks, Zach, 1985- author.
Description: Includes index.
Identifiers: Canadiana (print) 20220437580 | Canadiana (ebook) 20220437610 | ISBN 9780865719842 (softcover) | ISBN 9781771423731 (EPUB) | ISBN 9781550927771 (PDF)
Subjects: LCSH: Tractors—Handbooks, manuals, etc. | LCSH: Farms, Small—Handbooks, manuals, etc. | LCSH: Agricultural productivity—Handbooks, manuals, etc. | LCGFT: Handbooks and manuals.
Classification: LCC TL233 .L64 2023 | DDC 629.225/2—dc23

Funded by the Government of Canada

Financé par le gouvernement du Canada

New Society Publishers' mission is to publish books that contribute in fundamental ways to building an ecologically sustainable and just society, and to do so with the least possible impact on the environment, in a manner that models this vision.

LIABILITY STATEMENT

The author and publisher hold no accountability for use of the equipment described in these pages by readers or anyone else. This book is not a replacement for reading the tractor and implement user manuals or for the common sense and specialized training necessary for safe and effective operation.

Praise for

The Two-Wheel Tractor Handbook

Jam-packed with great information. This book is going to save serious growers a lot of time tinkering around.

— Curtis Stone, author, *The Urban Farmer*

When growing vegetables for market, one of the most important lessons is that bigger is not better. It's quite the contrary—even one acre of cultivated land can allow a grower to make a living in farming. Another lesson is to size the equipment properly, which is why two-wheel tractors have become so popular in recent years. In this fantastic book, you'll learn everything you need to know about this appropriate technology. This book is truly unique, full of tips, hints, and tidbits even for someone like me who's been farming with such tractors for over 20 years.

—Jean-Martin Fortier, author, *The Market Gardener*

Zach Loeks' *The Two-Wheel Tractor Handbook* is a detailed, beautiful, and much-needed must-have for every market gardener's library—and for anyone serious about using small machines to manage their farm.

—Matt Powers, author, *Regenerative Soil*

This book is essential to anyone who owns a BCS tractor or who is exploring the purchase of one. I use a BCS in the garden, forest, and the snow and found a ton of insights in Zach's book. The review of various BCS implements, different uses for these machines, and maintenance tips is invaluable to the novice and intermediate user alike. This book should be included as a how-to manual with every BCS tractor sold.

—Rob Avis, Verge Permaculture, co-author,
Building Your Permaculture Property and *Essential Rainwater Harvesting*

Easily the most comprehensive guide to two-wheel tractors ever written. Get more out of your two-wheel tractor with all the tips and tricks while saving money by doing the maintenance yourself.

—Diego Footer, founder, Paperpot Co.

As much as I favor managing your garden with hand tools, if you scale up to marketing you may need some help. In addition to serving as an operator's manual for two-wheel tractors, this book helps you define your goals for your enterprise and decide just what equipment you may need. A good resource to have.

—Cindy Conner, author, *Grow a Sustainable Diet* and *Homegrown Flax and Cotton*

This very practical manual will help us really understand our two-wheel tractors and get the best out of them while giving them our best. Even those in favor of minimizing tilling understand that some circumstances call for tilling, so let's do it well. Mindful, good use of machinery is important!

—Pam Dawling, teacher, speaker, and author, *Sustainable Market Farming* and *The Year-Round Hoophouse*

The scale of farming is crucial in designing farm systems and planning for labor, investments, profits, and machinery. This book discusses scale and choosing a family of machinery that work together and allow a farm to grow over time in a comprehensive and innovative way that I have not seen anywhere else.

—Sam Oschwald Tilton, direct market vegetable specialist, author of numerous articles and resources on sustainable agriculture, and vegetable farming mechanization specialist

TABLE OF CONTENTS

PRAISE

For Dad, thanks for keeping our hearts on track to a more beautiful world. Thanks to the many people who shared their stories and wisdom. Special thanks to Ryan, Stefano & Lorenzo for helping set up the Italian farm tours and to Larry, Vera and Ted for reading my drafts with scrutiny. This book's first illustrations were done almost five years ago, thanks to the journey and the publishing team at New Society Publishers.

PREAMBLE

Hey All,
Two-wheel tractors are great for small-scale growers! They have been around for a long time and really deserve much wider use across many land-based industries. The best implements are chosen to complete whole operation cycles—like building and preparing garden beds from scratch—and organized into equipment guilds as a practical design concept. Your projects and enterprises should scale up by setting a goal for an intended static scale... don't just keep getting bigger!

By setting goals for an intended static scale and understanding the principles of scale, growers can better select and use two-wheel tractors and equipment. Focusing on multi-functional equipment use and understanding enterprise profit and resilience is key to wise investment. I am a strong believer in investing in *technology, soil,* and *biodiversity.*

Grow on!

— Zach

INTRODUCTION

THIS BOOK'S FORMAT

This book is for current and future operators, covering background and basics as well as advanced topics, such as the Permabed System, and innovative techniques like the Compost-a-Path Method.

The holistic discussion covers principles to help growers adopt and adapt the presented techniques and designs to their own context, rather than having to apply a cookie-cutter plan.

This illustrated guide has hand-drawn designs by me, alongside photos, to help bring concepts together. Unless otherwise stated, I took all the photos. There are also full-page infographics (figures) covering the main design concepts.

The book's jargon has many terms for innovative two-wheel tractor use. Important terms will be in bold and defined briefly the first time they occur.

This book is organized into main topic chapters which are dotted with Pro Tips, Design Boxes, and Farm Features to help the conversation go deeper with case studies and examples.

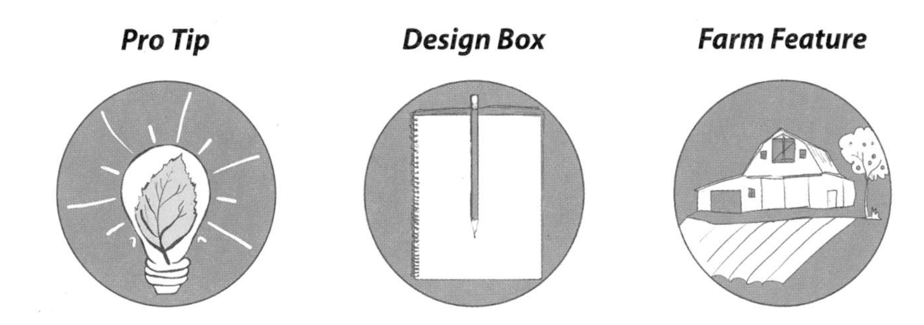

ABOUT THIS BOOK

Despite increasingly widespread *two-wheel tractor* use, agriculture and land-scape industries remain *defined* by four-wheel tractors. In part, this book's goal is to return the two-wheel tractor to its rightful place as a small-scale solution for land management, especially for diversified and highly profit-able stewardship of farms, homesteads, and landscapes.

Chapter 1 explores 100 years of two-wheel *tractor origin*, innovation, and renaissance to whet our appetite for broader and more innovative uses. This includes my own evolving experience.

Chapter 2 discusses this tractor's benefits, types, *essential components*, and operation.

Chapter 3 examines some *specific equipment* use, accessories, and adjustments.

Chapter 4 teaches decision-making for scale-suitable equipment using principles of scale, goal setting, equipment design, and enterprise planning. The chapter includes brainstorming exercises to help growers plan for the proper scale at different points in their farm evolution: ***start-up***, ***scale-up***, and ***pro-up*** scale phases.

Chapter 5 showcases different *enterprise types* with tractor, equipment, and operation recommendations to help growers situate themselves and incorpo-rate the design concepts discussed.

Chapter 6 outlines the ***Permabed System*** as a longer case study of equipment selection, adjustment, and use. Here innovative techniques, like creating Compost-a-Paths, are detailed for profitable and diversified land use.

Chapter 7 shows step-by-step equipment tasks to *transition any piece of land* to a garden system.

Chapter 8 helps growers achieve long-term success with key tips and exam-ples of *tractor maintenance and care.*

This book wraps up with an eye to the future of two-wheel tractors, crucial next steps in equipment design, and a call to action for change-makers and the future of food.

SMALL-SCALE PROFIT RESILIENCE

Hey, have you seen the potential of two-wheel tractors? This piece of equipment is small-plot maneuverable, start-up budget affordable, multi-enterprise functional, and future-need adaptable. Two-wheelers are increasingly being used by intensive market growers and homesteads, but what about using them in landscaping, orchards, and edible ecosystems? Two-wheel tractors are for *all* small-scale, highly productive, and **profit-resilient** land management! And by profitable, I mean in both the short- and long-term because they can be investments in soil, ecosystem services, biodiversity, and social capital. Enterprises that employ two-wheelers tend to be *profitable* (more income than expense) and *resilient* (nimble for socio-economic and environmental change).

Small-scale is often associated with *acreage-scale*. Fifty acres of mixed land use is small compared to 1,000 acres in wheat, but a 3-acre market garden or ¼-acre urban farm is even smaller! Small-scale also means *equipment-scale*; the equipment you use is part of defining the scale of your operation. Growers who can manage their land with only hand tools and/or two-wheel tractors have a smaller equipment-scale than those using primarily four-wheel tractors. Yet, small-scale doesn't mean lower profit or productivity! Profitable land management often contradicts "bigger is better"; intensive growers can make more income per acre with less land and equipment. However, larger-scale growers can also be quite profitable when efficiencies of mechanization and the ability to meet more needs in situ (such as growing your own mulch, fertility, etc.) become real savings and offer new profit centers. Resilience is usually achieved with a balance of scale; your acreage-scale and equipment-scale should be in harmony to suit a business and management model for your land.

Finding a profitable niche for your land, goals, and enterprises is helped by balancing **holistic scale principles** (see Figure 1), giving insight for *decision-making* no matter what your acreage, equipment, or production is. There is no stark boundary between large and small-scale success, and a shift in one principle, like actual production acreage, will change others: labor dynamics, equipment, or profit potential. For instance, a farm with sloping topography will be better managed with a method suited to that terrain. This will affect equipment and other aspects of scale, helping the farm meet its goals and achieve *steady-state profitable management* at a **static scale**. Your static scale is the intended goal of your operation's scale—that maximum of acreage, equipment, and other principles. Once your intended acreage is reached, for instance, you shouldn't acquire more land. You are at your static scale, and you can now operate and improve practices within steady-state management rhythms without further expansion.

That being said, two-wheel tractors and associated high-production methods are *usually* most suited to growing areas of ¼ acre (intensive urban gardens) to 1 acre (typical market gardens) but can be used for areas up to 6 to 13 acres (ex: agro-forestry). As such, this book discusses different tractor users with examples of the acreage, equipment, and production scales they might use.

This farm is making use of small fields using scale-suitable equipment.

Pro Tip: *Equipment decision-making is important because short- and long-term equipment selection is a major way that growers can make either the biggest mistakes or have the greatest successes. Having the right equipment can revolutionize your homestead or farm. The wrong equipment choice can dictate how you grow instead of facilitating how you want to grow.*

FIGURE 1: HOLISTIC PRINCIPLES OF SCALE

Holistic principles of scale are a shifting ratio of land investment and management. They form a decision-making matrix for growers to use to set goals, make plans, create designs, and run their operations. A growers goal should be a steady-state profit resilient land management at an intended scale.

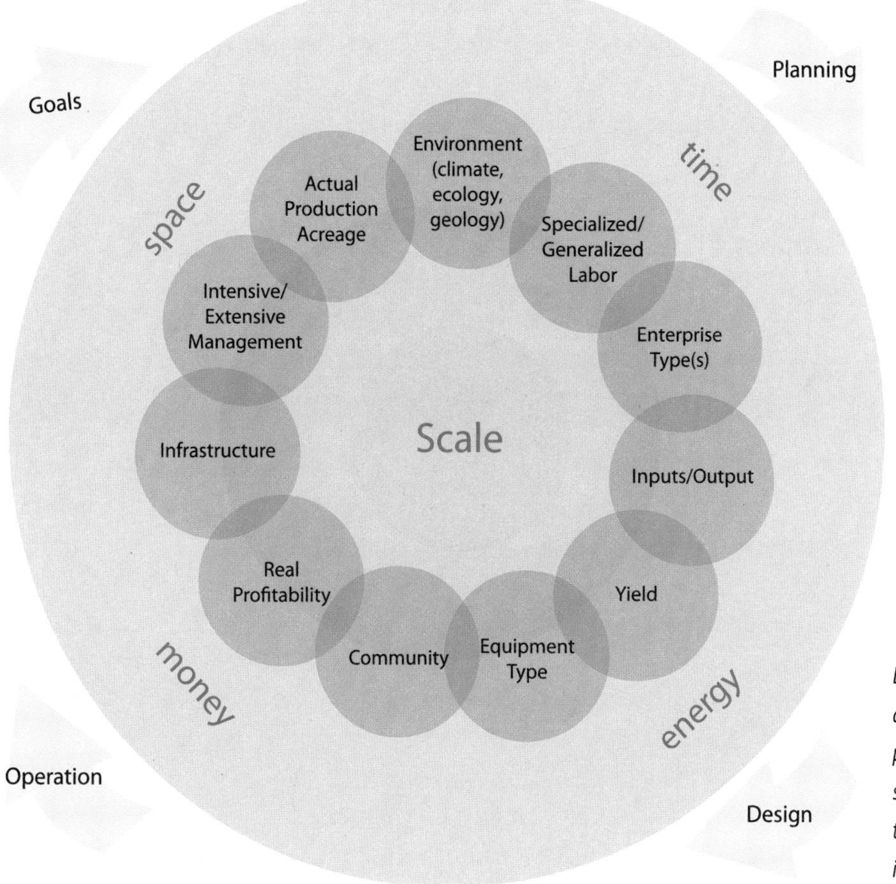

Decision-making is anchored in scale principles. These shift with space, time, energy, and the amount of money invested by growers.

Keep this matrix in mind and refer to it throughout the book and as you plan over the years to come.

Chapter 1
A Story About a Horse

The two-wheel tractor has come in and out of popular use for over 100 years. Its long history is similar to my own shortened history through the many stages of my agrarian business evolution. Let's explore the history of two-wheel tractors and my own connection to this small but mighty piece of equipment.

> ***"If you can't ride, can you fall?" (said the Horse)***
> ***"I suppose anyone can fall," (said the Boy)***
> ***"I mean can you fall and get up again without crying and mount again and fall again and yet not be afraid of falling?"***
>
> — C.S. Lewis, *The Horse and His Boy*

THE HORSE AND HIS BOY

I remember seeing that 10-horsepower "Italian Stallion" (as we fondly referred to that blue Italian-made, two-wheel tractor) for the first time one summer when I was about twelve. I was reading a book by C.S. Lewis under an apricot tree watching the Stallion open new ground for a three-sisters planting at our homestead in New Mexico. Like most kids, I was attracted to equipment, and this was a pretty neat machine! It had different attachments, looked like a little race car, and had the power to move earth in amazing ways. "Small but mighty," I thought.

What I liked most were the green rows of squash, salad, and peppers that grew in the tidy beds that the tractor formed. We were just one street inside the city limits, so the small stature of the Stallion was scale-suitable for our homestead—a property that was a back-to-the-land paradise, with gardens, orchards, greenhouses, small livestock, and more. My dad, an early permaculture adopter and designer, had been working part-time on our edible yards between landscaping and teaching jobs, and we kids would often tag along to help.

The Stallion was also great for the mountain farms in northern New Mexico that quickly adopted small-scale solutions for organic production in narrow agrarian valleys with *acequias* (traditional irrigation ditches that carry mountain stream water into the farm plots). One of these was a cooperative farm my brother and his friends started. The time I spent there in garden plots that followed coyote-willow streams has left lasting memories: the taste of pinyon nuts, the smell of Ponderosa pine, the colors of sage, and the feel of a cold-water creek—along with the jingle of the "pirate's gold," those half dollar coins Dad kept in his pocket as a reward for fully submersing ourselves. Yet, it was heirloom squash harvests, fresh potting soil, warm hoop houses, and the feel of steel turning soil that captured my heart and made me a lifelong gardener, homesteader, and farmer. "Small is beautiful," according to E.F. Schumacher, whose books I leafed through between landscape jobs.

When I started my own market garden in Ontario (where acequias ran as large as the Ottawa River), I used two-wheel tractor power to grow a rich variety of food. As the farm scaled-up over time, I integrated four-wheel tractors as part of what became a 10-acre permaculture market garden growing crops, cover crops, berries, herbs, and fruits for Community Supported Agriculture (CSA), farmers markets, and on-farm events. I still used the two-wheel tractor. It has had continued use for specialized jobs in greenhouses and the 100 ft caterpillar tunnels used for heat-loving tomatoes, peppers, eggplant, basil, and other high-value crops.

Now that my farm has transitioned production to become an edibleplant nursery supplying transplants for my company, **Edible Eco-system Design**, the two-wheel tractor has renewed importance. It has revolutionized my ability to maintain diverse edible plants efficiently and hit the ground running on an edible landscape installation! My dedication to *small is mightily beautiful* continues through research and education as part of **The Ecosystem Solution Institute**. I haven't forgotten about the horse and his boy as I prepare new food-forest beds.

Above and left: *Shown here are designs that won three Agri-innovation awards for the solar-powered farm, Permabed System, and a geothermal and ice-cooled cold storage/root cellar. These were used to farm organic vegetables and fruits, heirloom seed garlic, and specialty winter crops.*

Right: *Two-wheel tractors are a critical part of the edible plant nursery and ecosystem landscaping. Here we are making new beds for a tree nursery from scratch. Learn more at www.zachloeks.com*

Pro Tip: *Two-wheel tractors are great for landscaping because they can work in tight spaces, accommodate versatile implements, and can fit (with implements) in a small 6′ x 12′ trailer for easy transport to jobs.*

The farm has expanded to include numerous satellite projects, which together form The Ecosystem Solution Institute, an organization dedicated to education, propagation, and inspiration that was formed to encourage edible diversity and solutions for growers at all scales. A 100-acre Edible Biodiversity Conservation Area is currently under construction where thousands of edible plants are being trialed and designed into practical guilds for farmers and home gardeners. Learn more at www.ecosystemsolutioninstitute.com

THE ORIGIN OF TWO-WHEEL TRACTORS

Need is the mother of innovation, and the first two-wheel tractor emerged in the early 20th century as farms sought new equipment solutions for growing needs. This innovation was a new mechanized option for farmers who primarily used horsepower. The design was a multi-functional, self-propelled machine guided by a walking operator or one who sat atop the implement. These early implements were the draft type, actually modified from horse farming. Interestingly, in the 21st century, modern farms that use actual horse power and growers using two-wheel tractors operate on similar scales.

The transition from horse power included steam-powered tractors before gas and diesel became the main fuel. But whether steam, gas, or diesel were used, the difference between four-wheels and two-wheels was less distinct than it is today. In fact, the two-wheels or four-wheels were simply replacing the popular two-horse or four-horse configuration used by early farmers. They were just variants of the same innovation: the tractor. This is in stark contrast to how two-wheel tractors are seen today—as *less than* a tractor.

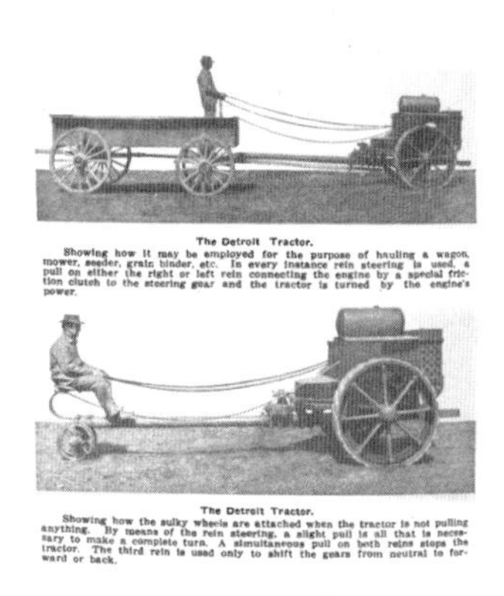

The Detroit Tractor.

Showing how it may be employed for the purpose of hauling a wagon, mower, seeder, grain binder, etc. In every instance rein steering is used, a pull on either the right or left rein connecting the engine by a special friction clutch to the steering gear and the tractor is turned by the engine's power.

The Detroit Tractor.

Showing how the sulky wheels are attached when the tractor is not pulling anything. By means of the rein steering, a slight pull is all that is necessary to make a complete turn. A simultaneous pull on both reins stops the tractor. The third rein is used only to shift the gears from neutral to forward or back.

In 1913, the Detroit Tractor Company introduced a "rein-driven tractor" reminiscent of early horse farming. Modern horse farming remains a good parallel for two-wheel tractor operations, which can handle a similar scale of operation. Photo source: Detroit Tractor Company, *Automobile Trade Journal*, July 1913.

Despite a century of continued two-wheel innovation, this tractor is all but forgotten on commercial farms. The four-wheel tractor spurred ahead as the powerful solution for the increasing farm size in America. Single-purpose power tools and small four-wheel machines became prevalent in property and landscape management as the two-wheel tractor was almost lost to sight. However, two-wheel tractors persisted where small land-holdings remained and intensive agriculture was valued. One such place, Italy, was and still is a hotbed of two-wheel tractor innovation.

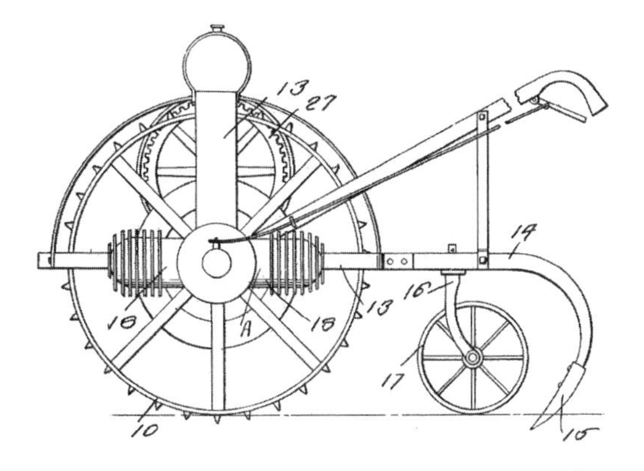

Early implements for two-wheel tractors included plows, cultivators, carts, and a new innovation: the rototiller. It was first patented by Konrad von Meyenburg in 1912 as a "Machine for Mechanical Tillage." In 1916, Benjamin Franklin Gravely innovated with his "Motor-Plow," which included an auxiliary motorcycle engine to help power a single-axle, single-wheel cultivator; he went on to be one of the lead innovators in American two-wheel tractor manufacturing. Photo Credit: B.F. Gravely, Jr., Motor Plow, US Patent 1,207,539, Dec. 15, 1916. and Konrad von Meyenburg, Machine for Mechanical Tillage, US Patent 1,016,843, Feb. 27, 1912.

HONORING THE TWO-WHEEL TRACTOR

Let's honor the two-wheel tractor as being a tractor in its own right. Equipment should be scale-suitable, and four wheels isn't synonymous with farming. The most important trend in agriculture today is arguably the pivot toward small-scale growing where small farming, permaculture, and ecosystem design will continue to be leaders in mitigating climate change and improving food security, safety, and sovereignty while enhancing local profitability.

THE ITALIAN TWO-WHEEL RENAISSANCE

In Europe, small fields are common. The farmland has been divided into smaller and smaller chunks over thousands of years. Walking the fields from Liguria down to Sicily, one can see that expansion was not as straightforward as it was on the great plains of North America, and perhaps that is a good thing. Need *does* bring innovation, and small *is* beautiful! In these small plots, the two-wheel tractor continued to be a practical piece of commercial farming equipment. Indeed, in Italy, the second half of the 20th century saw an expansion of two-wheel tractor use and innovation, which was then fed *back* to America. This all piqued my curiosity and sent me back to the source to see why two-wheel tractors had such a strong presence in Italy.

The country of Italy is shaped like a foot pressing down into the Mediterranean. Some people see it as a stylish Milanese shoe; sports fans see a cleat kicking a Sicilian soccer ball. But when you take into account that almost the entire breadth of the country is mountains covered by small-scale farms, it starts to feel more like a work boot! And work they did, in a land where a rich history spans prehistoric, Etruscan, Roman, and Italian cultures.

IMAGE CREDIT: ISTOCK-1213581408, PETER HERMES FURIAN.

The food culture of Italy is as diverse as the microclimates in those hills and mountains. One can find small fruit and vegetable greenhouses in the northern Dolomite mountains, hazelnut orchards in eastern Piedmont, and cliff-clinging homesteads and farms throughout the Ligurian Hills, including the famous basil terraces near Genoa. There are tomatoes, grapes, and apricots growing in deep soils in Campania under the watchful eye of Vesuvius and truffle-dotted woodland hills in Toscano amidst the great vineyards of Chianti. From Sicily to Sardinia and Naples to Genoa, the land is all mountains. Save, one area!

There are great open farmlands full of vegetables, rice, and grains in the northern provinces of Lombardy, Emilia-Romagna, and Veneto. These large, open fields are one of the points where mechanization emerged in Italy at the turn of the 20th century, and it was here that the two-wheel tractor became an essential part of Italian farming. Yet, it was the rest of the provinces, with their immense hilly terrain and small fields, where the two-wheel tractor remained most relevant.

Here, in the Italian Dolomites, the two-wheel tractor is scale-suitable to small brassica and radicchio fields and narrow high tunnels growing tomatoes and salad. The diversity of food at an Italian farmers market is staggering. The land and its micro-climates allow a plethora of productions, and the people know how to cook and enjoy this diversity. This all supports the small-farm growth that is returning to the Italian countryside, reinvigorated by the new food revolution! This new revolution includes a return to living off the land, growing food for local and regional markets, and reinvesting in scale-appropriate technology innovation.

FARM FEATURE

THE CASTOLDI STORY

Cesare Castoldi grew up on a dairy farm near Abbiategrasso, Italy. Like most early 20th-century farms, they had gardens, hay, and orchards. Cesare was one of four children—all boys. Back then it was a long horse carriage journey to town, so the children spent their days on the farm playing and doing chores by hand—but young Cesare had an eye to the future!

One day, a traveling salesman stopped at the farm and remarked that the boys were of the right age and that "not far from here, only 20 km, there is a family with five girls. You are all nice and well-educated," he said, with a match-making smile. Cesare and Adele began to court, and they were happy and did marry in 1895. Their third child was named Luigi Castoldi, who became a key contributor to the small-farm revolution we have today, though he didn't know it then, and it would take many years and many innovations to make it happen.

Cesare and Adele were a smart match, and she was a very smart lady. Cesare was now running the fields of the family farm, and Adele managed the business. In 1929, just before the Great Depression, she proposed that they withdraw all their savings from the bank to buy another big farm. When the stock market collapse came, the effect on the Castoldi farm was lessened because they had put their earnings into land. This kept their dreams alive alongside the aspirations of their children.

Young Luigi Castoldi had a "strong attitude for the design of machines," his son Fabrizio remarked many years later, and Luigi's brother Achille also loved mechanics and sports. Achille went on to win gold as an Olympic rower in Warsaw, Poland. When the family went to the farmers market, everyone congratulated Cesare on his family's sporting and academic achievements. Little did he know that a unique family enterprise was developing.

Cesare was an early adopter of technology, including a recently imported Fordson tractor. So, in celebration of the gold medal, he bought Achille a modern mechanical wonder: a motorcycle! Achille was proud of his motorcycle, a local first, and he drove it everywhere, startling people and horses alike (none of whom had ever *heard* of one before). Luigi asked constantly if he could drive his older brother's motorcycle. Achille said, "No!" But Luigi wasn't one to give up easily, and when he was refused for the umpteenth time, he asked Achille if he could simply *clean* it. Achille agreed.

The excuse to clean the motorcycle granted, what Luigi *actually* did was disassemble it completely in the living room (with his younger brother Aldo polishing each and every component). To put it mildly, Luigi was extremely fond of the inner workings of machines.

By 1930, Achille had placed on the podium in 70 races, and he was getting too old for rowing. He still loved water sports and decided to try motorboat racing and wanted to enter with a bang. So he talked with his brother Luigi about an idea. At the time, the popular Johnson speedboat engines

Luigi cleaning his brother's motorcycle—as described by his son, Fabrizio Castoldi—who sketched out for me some of the early innovations his father made that led to the founding of a two-wheel tractor company.

Image Credits: Courtesy of the Castoldi Family and BCS

(300cc and 500cc) were American-made and often suffered malfunctions as a result of being transported across the ocean. The European engines were not powerful enough to compete in these races, so the Castoldi brothers needed a solution that would allow the use of Italian engines that would also give them a competitive edge.

So, Luigi told his brother, "We need to work on this!" Then, with cousin Mario, a designer of early water-landing planes, they sketched out a new race boat for Achille. Luigi and Mario's design, built by local boat builders, was narrow, with two outrunner wings and triangle-fabrications underneath. When the boat reached speed, the hull would rise, lifted by the wings, leaving only three points of the hull touching the water. With this early hydroplane solution, Achille started racing and won a world record on the channel outside of Milan. In the next race, on the same day, he won another world title. Two world records in one day. This is a story of engineers and an athlete!

Luigi had been studying engineering in 1924 in Turin, but when he took a break and returned to farm with his father, he stayed for six years. It was during this period the speed boat was innovated, reinvigorating his interest and spurring him to finish his degree. Luigi started turning his engineer's mind toward many agriculture projects, including helping to develop a micro-hydro project for a local farm's electrical lighting.

Then, knowing well the arduous work of cutting hay by scythe, Luigi innovated a hay mowing machine. In 1940, his prototype was built in the horse stables on the family farm and released for sale in 1942. The horses looked at him askance. Luigi had begun a journey developing a new form of *horse power* for local farms! Today, Luigi's innovations live on and "Castoldi" is the "C" in BCS, the popular two-wheel tractor company that makes the tractors found today in market gardens, and on homesteads and farms—and the *Italian Stallion* on our own small-scale farm in New Mexico.

Engineer Luigi Castoldi personally tests his invention in 1942: the new Motofalciatrice *243.*
PHOTO CREDIT: FABRIZIO CASTOLDI

AMERICAN RETURN TO SMALL FARMING

In America, the two-wheel tractor has been picked up again by growers whose success from *direct marketing* makes small-scale farming profitable. For 50 years, the market gardening revolution and a back-to-the-land movement has adopted two-wheel tractors as their equipment-scale.

There are more than just Italian influences on the new food revolution in North America. Traditional French market gardening and Japanese **Teikei** met forward-thinking Northeastern growers like Eliot Coleman, who innovated techniques and tools for market gardening and started farming in the

1960s, and Robyn Van En, who started the first **Community Supported Agriculture (CSA)** farm in 1985. Books by both those innovators were important for me as I started up my own market garden and CSA. At the same time, Indigenous Australian knowledge was melding with ecological science into a new design methodology called **Permaculture**, which was emerging in the 1970s due to innovative thinkers like Bill Mollison and David Holmgren, who were finding fertile ground in techniques that encouraged a more regenerative society and agriculture. Permaculture caught on quickly in the Western states; one of the hot spots was New Mexico, where my dad was a designer in the 80s and 90s.

These movements and pioneers helped adopt and adapt small-farming to the modern context. Two-wheel tractors became the equipment of choice for a new wave of organic growers at the end of the 20th century. This pace has only increased with the turn of the 21st century. I started my market garden in 2007, purchasing a BCS 853, which had great improvements and additional features compared to the 725 I grew up with.

THANKS GIVING

Scale-suitable tools, techniques, and equipment are essential to land profitability and resilience. Ideas have been shared across oceans, from ancestral lineages, and always with an eye for reinvigorating innovations in new contexts. Let's give thanks to all early food innovators, and primarily the Indigenous peoples of the world who tended the first soils upon which our ideas and innovations grow.

Chapter 2
Two-Wheel Tractor Essentials

Let's now define the two-wheel tractor and consider its many benefits. Then, we will explore the different types of tractors and implements. After that, we will lay out some decision-making paths that will be useful when selecting a tractor for the first time—or when updating an older model.

THE DEFINITIVE TWO-WHEEL TRACTOR

This tractor goes by many names: "two-wheel," "single-axle," and "walk-behind," all of which highlight its small-scale nature. But the word "tractor" defines its *multi*-functional use and sets it apart from the plethora of look-alike single-purpose power equipment. It has one system (engine, transmission, wheels, PTO) powering multiple interchangeable implements. It can do the work of a shed-full of power equipment, with only a single system to maintain.

A tractor can be simply defined as a land-management machine that has a universal hitch and PTO so it can operate various implements to perform different jobs—from mowing to tilling, and snow blowing to hay baling. A **PTO (Power Take-Off)** is an efficient mechanical part that carries the energy transferred along the drivetrain from the engine and applies it to any number of implements that have no power of their own. Innovated in the first half of the 20th century, the PTO revolutionized farming by allowing a wide range of implements to be developed, all of which are engaged by the mighty power of an engine rather than relying on being ground-driven (like a compost spreader), or pulled through the soil (like a plow, which usually has no moving parts). Whereas, a PTO-powered implement (like a roto-tiller) is powered by the PTO connection to a tractor's engine. We will look more at different types of implements later. Many two-wheel tractors have PTOs, and all have hitches for multiple implements.

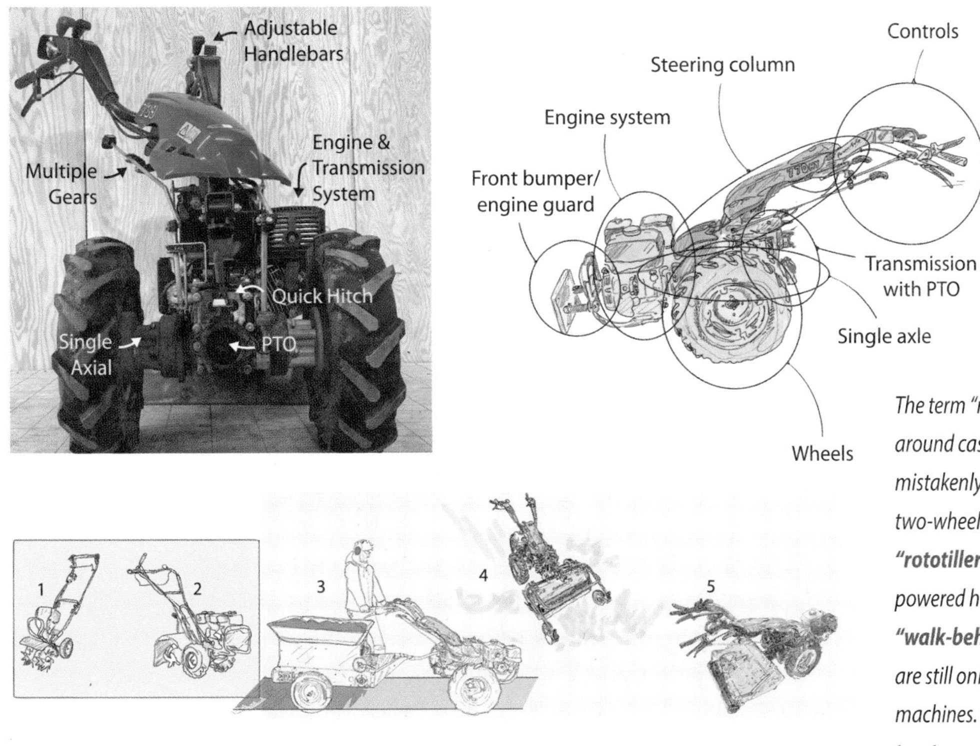

The term "rototiller" is thrown around casually and is often mistakenly used to refer to a two-wheel tractor! (1) Small ***"rototillers"*** *are essentially powered hand tools. (2) Larger* ***"walk-behind rototillers"*** *are still only single-implement machines. (3)* ***Two-wheel tractors*** *are in a class of their own. They have a PTO and hitch system similar to four-wheel tractors, and they can connect to multiple implements. Their versatile design allows the ability to rotate the handlebars, which allows the two-wheel tractor to be used in two different implement modes: (4)* ***front-mounted*** *(implements like flail mowers), or (5)* ***rear-mounted*** *(implements like rear-tine tillers with precision depth rollers). More on types of implements later.*

ELEVEN TWO-WHEEL TRACTOR BENEFITS

Two-wheel tractors are great small-scale equipment, making land management a breeze. They are suitable for many enterprises that need affordable, maneuverable, multi-functional equipment.

1. **Multi-functional:** Two-wheel tractors have only one engine, yet multiple implements can be used—opening up opportunities for many enterprises.
2. **Easily learned:** These tractors have a simple but effective design. They have a short learning curve compared to larger equipment.
3. **Task-appropriate power:** Most two-wheel models have between 4 and 13 hp, but some can be found as low at 1.5 hp and some upward of 16 hp. This is a practical power range for many small-scale jobs that are often done with overkill engines in landscaping and farming.

4. **Budget-friendly:** A grower can pick up a tractor and all needed equipment on a modest start-up budget. Two-wheel tractors start at around $2,500, whereas the smallest "compact" four-wheel tractors still cost ~$13,000 to ~$16,000. Two-wheel tractor implements cost between $500 to $3,000, compared to $5,000 to $15,000 for four-wheel tractors. ***Note:*** All prices are in USD as of 5/2022.
5. **Seasonal implements:** The two-wheel tractor performs (with the correct equipment) spring, summer, fall, and winter tasks.
6. **Equipment options:** Equipment comes in different widths to match the scale of your operation and with different accessories that can be customized to your terrain and tasks. Different equipment versions suit different needs and budgets. For example, there are more than three types of plows; each has its own merits.
7. **Low impact:** Two-wheel tractors cause far less compaction on your soil and use less fuel—while still doing a job right.
8. **Low maintenance:** Two-wheel tractor maintenance is straightforward. Components are easily visible and accessible, and there is no need for specialized tools beyond those found in a typical home garage.
9. **Maneuverable and easily controlled:** These tractors are maneuverable, well-balanced, small, and have a tight turn radius. This is ideal for negotiating sloped land, garden headlands, and greenhouses. There are two main types of two-wheel tractors (discussed shortly), and both are very maneuverable. The first type, the *row crop tractor* provides excellent cultivation and seeding control with its great hitch design, while the most popular type, the *multi-purpose tractor*, has a drive system that is perfect for maneuvering with loaded carts, mowing, and heavy soil working.
10. **Easy storage and transport:** These machines are easy to store. With foldable handlebars and a compact size, they can be loaded in the back of a pickup truck or on a small trailer.
11. **Safe handling:** Their scale and features make them safe for operators at all levels of experience (with proper training, of course).

THE TYPES OF TWO-WHEEL TRACTORS

There are a few variations of two-wheel tractors to consider. These can generally be placed into two categories: **multi-functional** (with PTO) and **row crop** (primarily for cultivation). Smaller models were popular in the 20th century, and they are returning for special applications.

Note: Unless otherwise specified as a four-wheel tractor, all tractors—both multi-functional and row crop—discussed in this book are the two-wheel type.

DESIGN BOX
SMALLER TWO-WHEEL TRACTOR

Small older tractors can still be found used, online, and at auctions, like the "Planet Jr Tuffy" and "Super Tuffy" models, which are similar to motorized wheel hoes or what was called a *moto-plow* in the early 20th century. These are still great tools and there are lots of choices out there. Tilmor has now come out with a new electric model, the E-Ox, which will be great for cultivation for backyard gardeners, intensive urban growers, and greenhouse applications for larger farmers.

ROW CROP TRACTOR WITH HIGH CLEARANCE

Row crop tractors are designed for crop row cultivation using a variety of hitch-attached implements. There are a number of *different cultivators* available for specific weeding applications, including finger weeders for in-row weeding or tender plant hoes for small crop management, which are assembled as a **gang** (arrangement) of cultivators on a toolbar to meet different crop needs as a whole unit suitable to the farmer's row spacing and crop needs. For small farms and serious Back-to-the-Landers, the row crop tractor provides mechanical weeding solutions for row vegetable production. These are the small cousin to four-wheel cultivating tractors like the Allis-G or Farmall 100.

The only new two-wheel row crop tractor in North America is the Power Ox by Tilmor, but you can still find used cultivating tractors like the Planet Jr BP-1. They both have great applications on a row crop farm. The Tilmor has great clearance (13.5") and a slew of new implements manufactured by Thiessen Equipment that work well with it. The implements are modular for easy adjustment to suit your row systems.

Note: Row crop tractors, like the Planet Jr and Power Ox, don't have a PTO and cannot power PTO-powered implements, nor do they have enough horsepower to pull plows or subsoilers. They are not meant for slopes or heavy-debris soil conditions.

Row crop tractors are specialized for cultivation to manage weeds and improve the soil/air interface, similar to earlier forms of organic agriculture when horses were used and synthetic herbicides didn't exist.

IMAGE CREDIT: S.L. ALLEN COMPANY

ROW CROP TRACTOR FEATURES

1. **Hi-clearance** means more under-tractor height, which allows crops to be cultivated without damaging them. Clearance is often achieved with a chain/belt drive system located beside the engine instead of lower down. Tractor clearances of 7" to 13.5" allow 15" to 27"-tall crops to bend underneath the axle during cultivation passes.
2. **Culti-vision** is the ability to see the crop you are weeding. Any two-wheel tractor provides great visibility, but some have better under-tractor visibility to aid cultivation of row crops.
3. **Lower hp engines** (~1–4 hp) are fuel efficient for light-duty cultivation (usually .5" to 1" deep for hoe weeding, deeper for hilling applications); deeper cultivation is *too late* and weeds are too big. More power is needed to plow, so you have to use different tractors types!
4. **A central implement hitch** is located near the center of the tractor axle so implements have a better response when the wheels are turned—essential for accurate cultivation.

5. **Adjustable handlebars** for height, but not reversible for using push-type implements (there are no commonly used push-type implements for row crop tractors).
6. **Implements are modular** for mounting and arranging different cultivators as a gang.

Above left: Tourne-sol Co-operative Farm team in front of greenhouse. Photo credit: Tourne-Sol Co-operative Farm

Above right: This great collection of Planet Jr cultivators helps William Staufer, a Back-to-the-Lander. At his farm's peak, he had about six acres of extensive vegetables, which he maintained mostly by himself, along with some seasonal planting and harvest help. Photo credit: William Staufer

Below: Reid Allaway at Tourne-Sol Farm talks cultivation options for his fixed-up Planet Jr BP-1, including tender plant hoes, finger weeders, and hillers. Photo credit: Ghislain Jutras – L'Odyssée Bio de Gigi

MULTI-FUNCTIONAL TRACTOR WITH PTO

Multi-functional two-wheel tractors (referred to in this book as **M2w-tractors**) are the most popular type of two-wheel tractor, with applications across many industries: farming, landscaping, property maintenance, forestry, orcharding, and market gardening. M2w-tractors can be used with multiple implements, including PTO-powered types. The innovation of rotating handlebars is a key multi-functional trait of M2w-tractors because they provide a typical rear-mount mode for pull-type implements as well as an alternate front-mount mode for push-type implements. Some of these implements are PTO-powered and others are non-PTO implements.

Note: You should never try to *push or pull* a tractor. The engine, wheels, and proper use of weights do this work for you!

FIGURE 2: M2W-TRACTOR

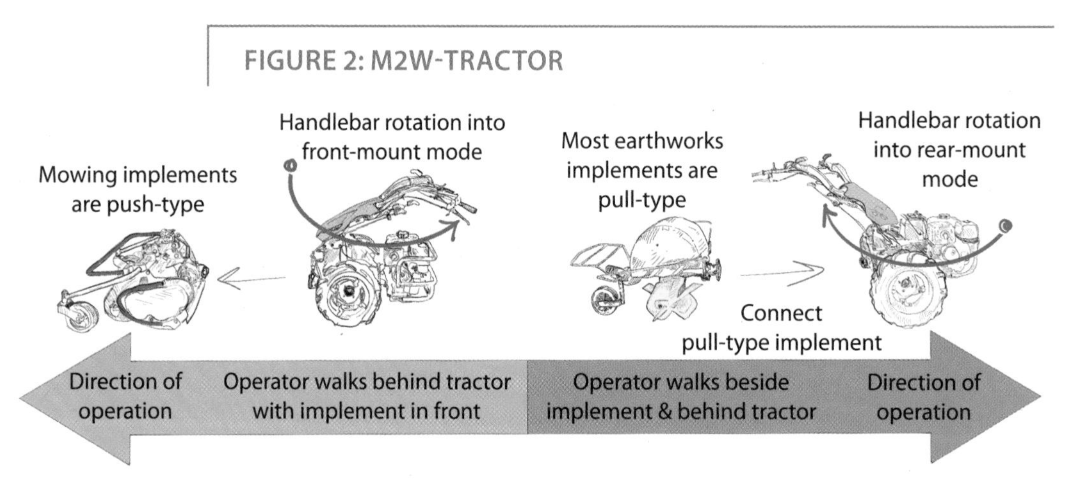

M2W-TRACTOR FEATURES

1. **This tractor has many uses** for different projects and enterprises since it is not limited to light-duty jobs, and it has *many* equipment options.
2. **Implement-type versatility** includes the ability to use both PTO-powered and non-PTO implements.
3. **Lower clearance** is less desirable for cultivation, yet better suited for earthworks and mowing-type jobs because of a *lower center of gravity.*
4. **Higher hp engine** (usually between 8 and 13 hp) provides the necessary power for a diversity of tasks.
5. **More transmission/clutch options** include mechanical, or hydro-mechanical, clutches and all-gear, or hydrostatic, transmissions.
6. **Adjustable and rotating handlebars** have anti-vibration mountings and improve ergonomics; operator height adjustments for operators and the offset handlebars improve operation in unique situations, like raised bed management.
7. **Rear- and front-mount modes** improve equipment options, transport, and general operation.

TRACTOR COMPONENTS: FORM AND FUNCTION

The two-wheel tractor can be organized into its **major components** and its **essential parts** to help operators make decisions about which optional features and accessories they need. The infographics shown here along with the index identify most two-wheel tractor parts and show three popular models (the BCS 739, 853, and 770), which have a range of sizes and features. These visuals also provide insight into other tractor types and models that you might have or choose to use. ***Note:*** The index numbers are used across all models to show similarities and differences. A features availability for a certain model is indicated in parentheses. Example: (Electric start models only).

Pro Tip: *With a solid understanding of a tractor's parts, we become familiar with their use, options, and maintenance.*

M2W-TRACTOR COMPONENT INDEX

Note: Figures 3, 4, and 5 show components of M2w-tractors. Figure 3 is the BCS 770, Figure 4 is the BCS 739, and Figure 5 is the BCS 853. Some components are similar to those on two-wheel row crop tractors. This section is meant to help you orient around the tractor as you are learning about components. I further elaborate on most components later.

1. **Recoil** (rope) and **Starter Handle**
2. 2a. **Engine Choke** 2b. **Fuel On/Off Lever. *Note:*** This is not pictured on the BCS 853.
3. **Recoil Starter Assembly**
4. 4a. **Fuel Tank** 4b. **Fuel Cap**
5. **Air Filter Assembly**
6. **Muffler Assembly**
7. **Engine Guard/Bumper. *Note:*** The guard is not pictured on the 853, but it comes with one.
8. **PTO Engagement Rod**
9. For hydrostatic models, like the 770, this is the **High/Low Wheel Speed *Range* Selector Rod.** For all other models, like the 739 and 853, this is the **Wheel Speed Selector Rod.**
10. Tractors with the **Electric Start** option will have a **Key Switch** on the side (10a on the 739). Older recoil start models included an **On/Off Switch** in the same spot

(10b on the 853, see Figure 5). PowerSafe models feature an **Engine Kill Switch** on the handlebars (15).

11. 11a. **Left Handlebar Grip** 11b. **Right Handlebar Grip**
12. **Clutch Lever**
13. **Operator Presence Control (OPC) Lever**
14. 14a and 14b. **Individual Wheel Brake Levers.** ***Note:*** Individual brakes are not available on model 739.
15. **Engine On/Off Switch** (PowerSafe models only)
16. **Parking Brake** (PowerSafe models only)
17. **Handlebar Rotation Control Lever** (for turning handlebars between front-mount and rear-mount mode)
18. For BCS 739, this is the **Shuttle Reverse Lever**. ***Note:*** Not available on BCS 770. See #20.
19. **Hand Throttle Lever**
20. For BCS 770, this is the **Hydrostatic Speed Control Lever** (forward/reverse and ground speed control)
21. **Differential Lockout Lever**
22. **Handlebar Height Adjustment Lever**
23. **Wheel/Tire**
24. **Wheel Extensions** (standard on models like the 770 or 779; optional accessory on all other models)
25. **Bumper Weight Kit** for 770. ***Note:*** This is an optional accessory.

26 26a. **Steering Column Cover/Shield**

26b. For recoil start models, like the featured BCS 770 and 853, this is the **Tool Box Cover/Shield**· ***Note:*** For electric start models, like the 739 featured here, this is where the battery goes and there is no cover/shield.

27 27a and 27b. **Handlebars, left and right.** ***Note:*** These are adjusted up and down with the #22 lever, and rotated left and right with the #17 lever.

28. **Battery.** This is for electric start models only. ***Note:*** The 739 featured here is the electric version of the 739 model.
29. **Engine Oil Dipstick** (only pictured on BCS 853). See the maintenance section (chapter 8) for more details.
30. **Attachment.** ***Note:*** A power ridger is pictured for BCS 770, a rotary plow for the BCS 739, and a 30" tiller for the 853.

FIGURE 3: COMPONENTS BCS 770

Featured here is a bird's-eye view of the BCS 770 in front-mount mode for a push-type implement like a power ridger. This tractor has a recoil start, hydrostatic transmission, PowerSafe clutch, and front bumper weight.

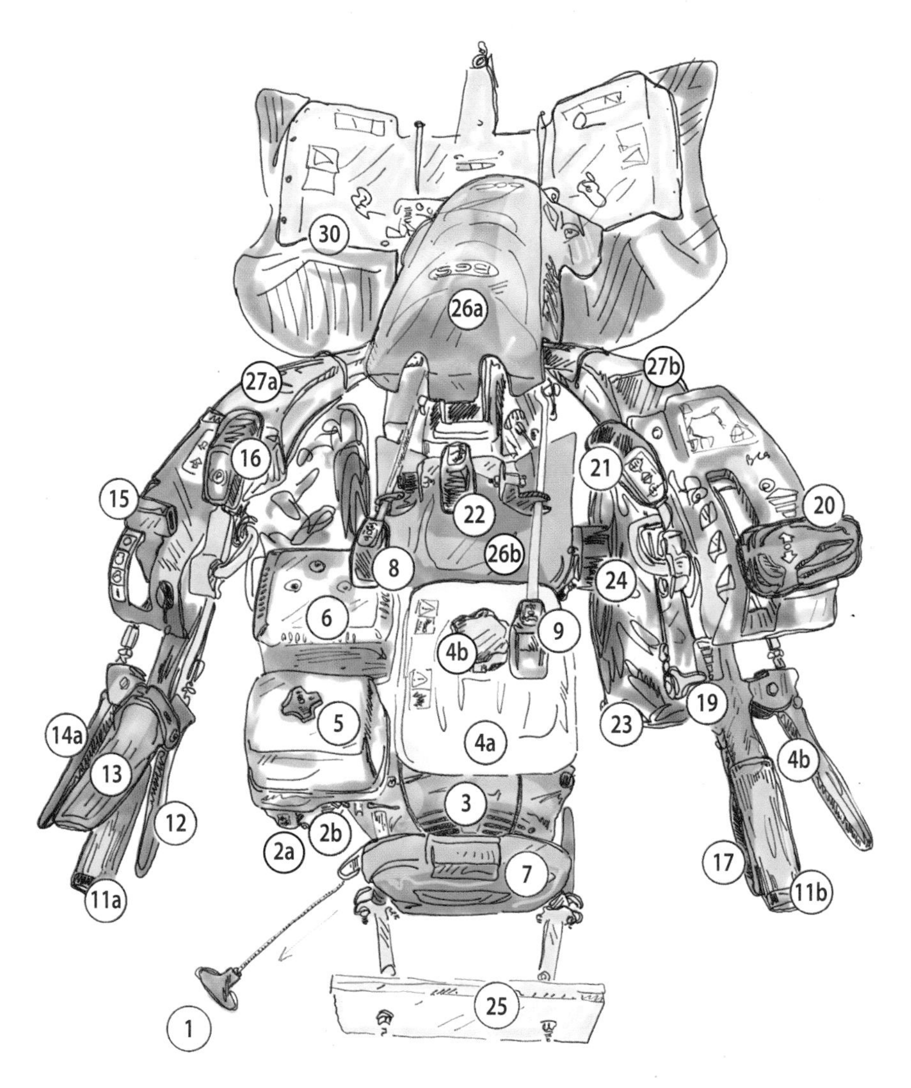

The hydrostatic drive is great for landscaping and mowing applications but is more than you need for gardens and market gardens.

FIGURE 4: COMPONENTS BCS 739

This BCS 739, in rear-mount mode with a rotary plow, has all-gear drive, an electric start option, and no right and left brakes. It has a hydro-mechanical clutch and a 3-speed transmission.

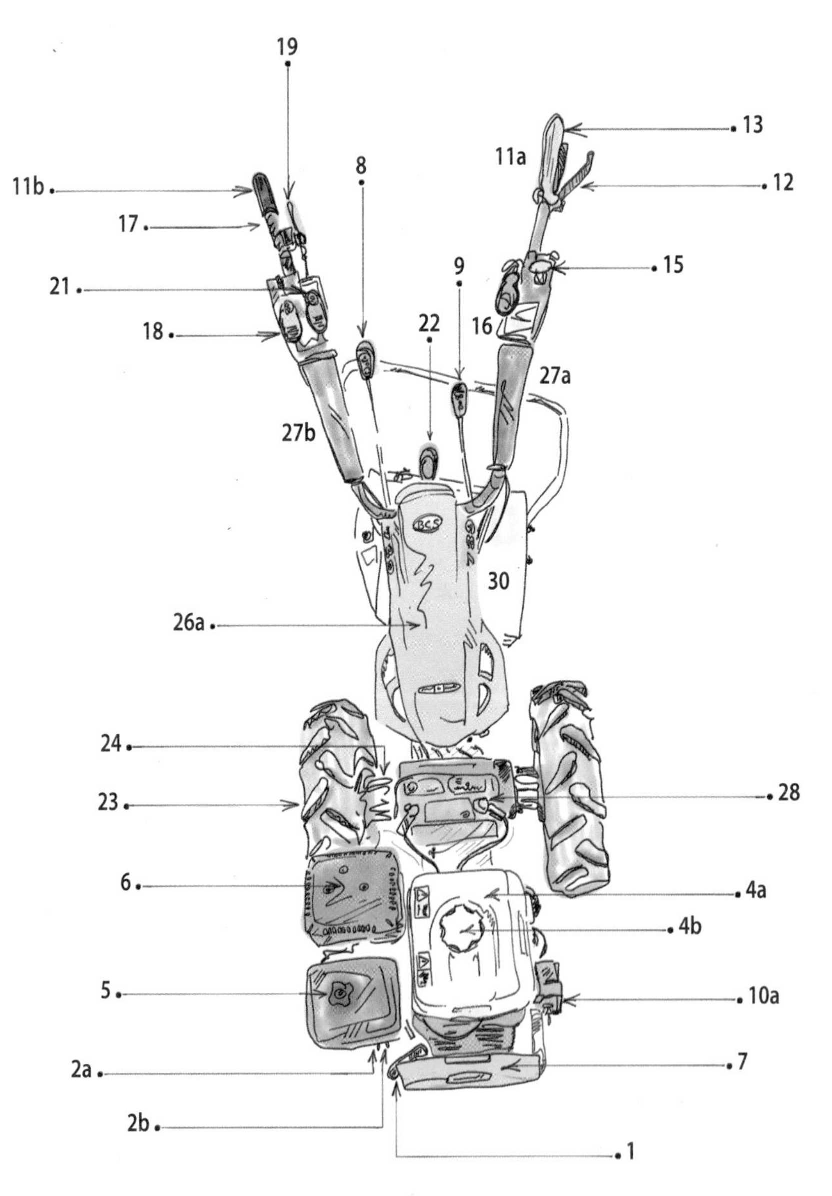

The electric start is easy to use, but battery maintenance is necessary for models with this feature.

FIGURE 5: COMPONENTS BCS 853

This BCS 853 is my old tractor. It is shown in rear-mount mode (these modes are discussed shortly) with a rear-tine tiller. It has an all-gear drive, a recoil start with an engine on/off switch, differential drive, and right/left braking. It has a double cone clutch and a 4-speed standard transmission, including a road speed.

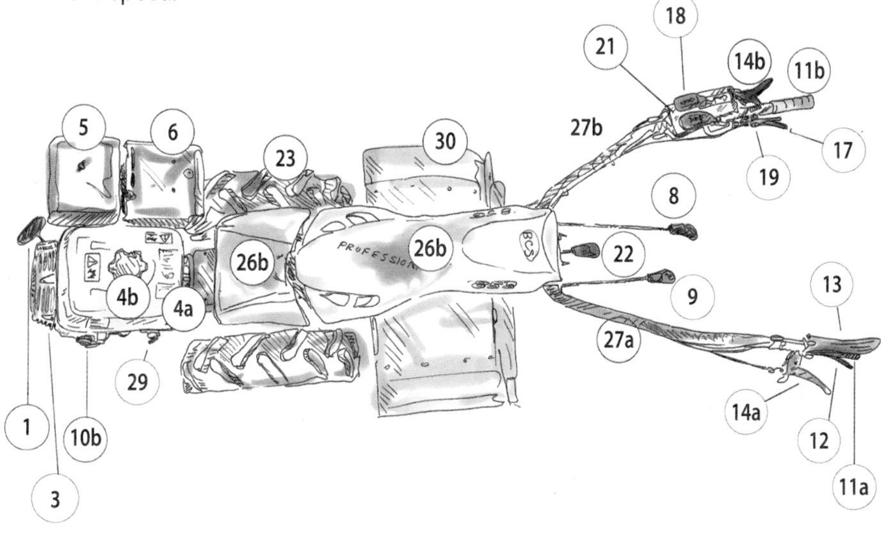

This standard transmission tractor features a differential drive, a must for garden applications where growers are often turning at the end of short rows; the differential drive is also available on other popular models with a PowerSafe clutch.

FIGURE 6: MAJOR COMPONENT FUNCTIONS AND OPTIONS

MAJOR TRACTOR COMPONENTS

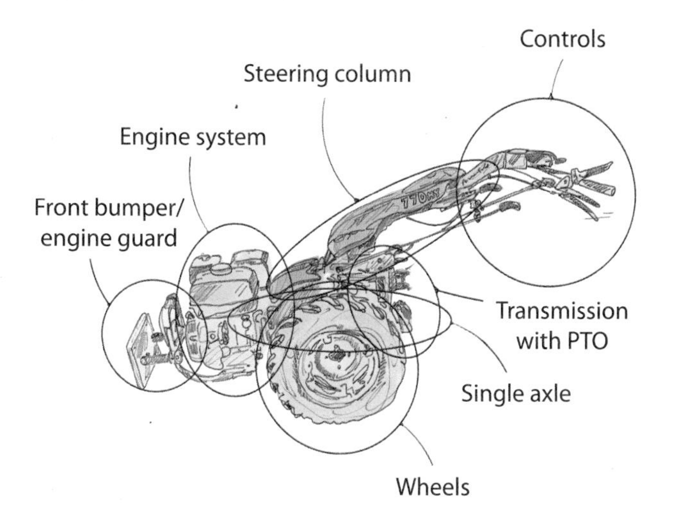

Tractor brands and models have different features, but the main components are the same.

THE FRONT-END

The front-end of the tractor includes the front bumper/engine guard and optional front weights. This area is designed to protect and provide additional weight for operating certain implements that require this.

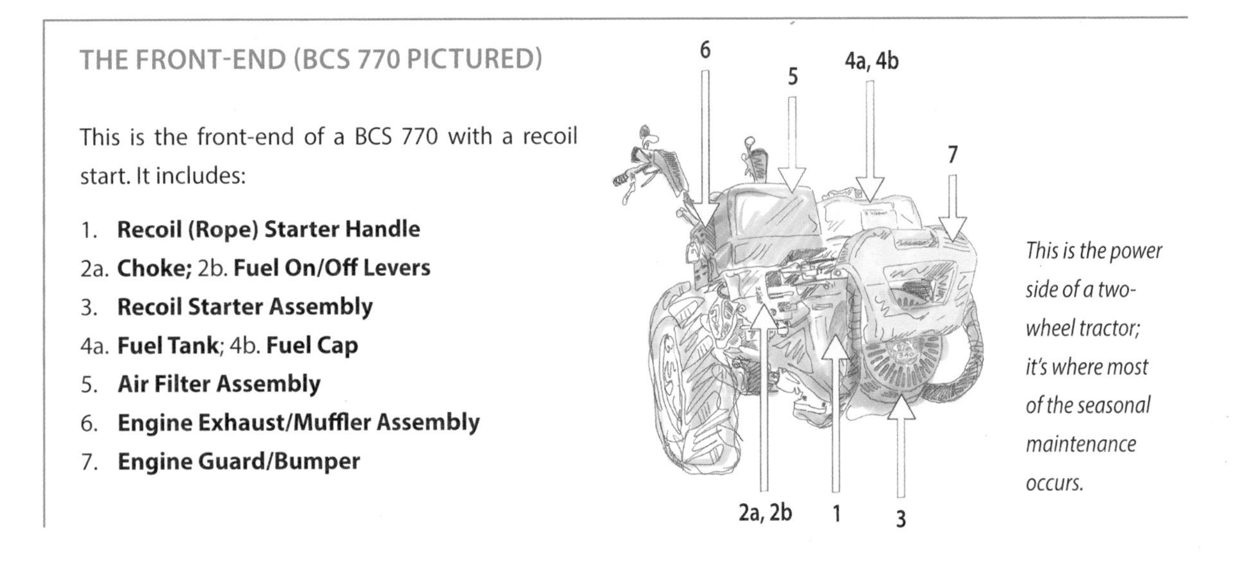

THE FRONT-END (BCS 770 PICTURED)

This is the front-end of a BCS 770 with a recoil start. It includes:

1. **Recoil (Rope) Starter Handle**
2a. **Choke;** 2b. **Fuel On/Off Levers**
3. **Recoil Starter Assembly**
4a. **Fuel Tank**; 4b. **Fuel Cap**
5. **Air Filter Assembly**
6. **Engine Exhaust/Muffler Assembly**
7. **Engine Guard/Bumper**

This is the power side of a two-wheel tractor; it's where most of the seasonal maintenance occurs.

ENGINE SYSTEM

The engine system is the powerhouse, including air intake and filtration, fuel tank, carburetor, combustion chamber, and starter system. ***But how much horsepower do you need?*** Professionals need more power to operate larger, heavier, and PTO-powered implements. Entry-level engines are great for backyard gardeners. The Honda GX390, BCS 853's standard engine, is an 11.7 hp (8.7 kW) 4-stroke engine, suitable for professional equipment like the 35" flail mower or 32" power harrow. Honda has been making engines since 1947. They are well-made and low maintenance, with readily available servicing and parts. They are also common in other farm equipment like water pumps and backup generators—simplifying your maintenance know-how!

Note: The engine system is set far forward on the tractor—in front of the wheels—to help balance the tractor. The two-wheel tractor is *not* balanced *without* an implement, but it *is* designed to be well-balanced when the engine is *counterweighted by an implement.*

WHAT DOES YOUR ENGINE SIZE DO?

- All engines have similar maximum speeds; faster ground speed is achieved with a larger tire diameter.
- The selection of engine size is part a factor of weight (as discussed), but it is also a question of horsepower.
- Engine power should be balanced by equipment load such that operation of the implement under normal working conditions won't drop the engine's rpm (revolutions per minute) significantly. Implement recommendations are based on both of these factors: engine power and front-end weight.
- More power is needed for heavier loads and increased traction (with proper tractor weights) in poor soil conditions.
- Higher torque helps in *getting started* when pulling a plow or subsoiler through the soil.
- Row crop tractors have lower hp engines because they mostly pull light cultivators.
- Tillers can be operated with smaller engines, but some PTO-powered equipment need larger engine horsepower. Examples include flails, brush and lawn mowers, snow throwers, and chipper/shredders.

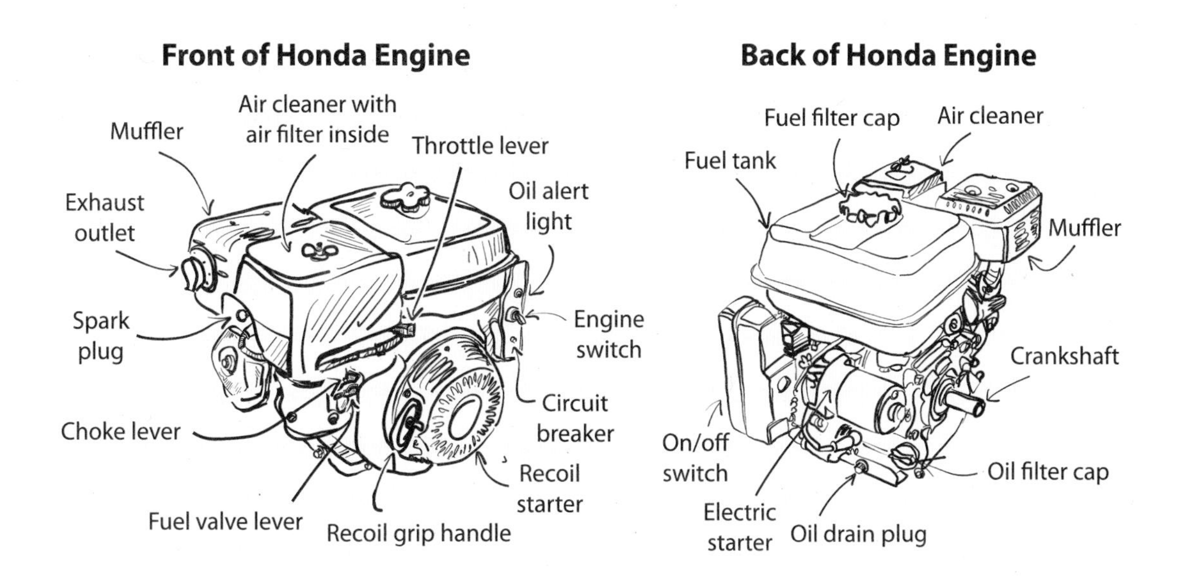

EQUIPMENT FEATURE

USED TRACTOR ENGINES

When buying a used tractor, consider that many engines have been popular over the years: Briggs & Stratton, Honda, Loncin, Kohler, Toro, Subaru, Yanmar, and ACME. When buying a used tractor, identify the make and model, and source needed parts ahead of time. It is often best to install a new engine. As of now, BCS uses mostly Honda and Kohler engines; Grillo uses Loncin, Briggs & Stratton, and Honda; and Tilmor uses a Honda engine. These are all good engines, and the parts are readily available.

Shown here are three views of an old ACME engine. When an old engine doesn't work right, identify the model number and the year, and replace it!

HANDLEBARS AND CONTROLS

The handlebars are used to steer, and they house the following key controls: differential lockout, clutching, and emergency shut-off, using the ***Operator Presence Control*** (OPC). The OPC stops the tractor's movement if the operator lets go of the handle—which is a good thing in an emergency situation. The controls for shifting gears and engaging the PTO are accessible below the handlebars. M2w-tractors also have adjustable handlebars and steering columns.

Pro Tip: *When idling in the field, shift into neutral before engaging the U-clip on non-PowerSafe models, otherwise the engine might shut down. Also, for storage, the U-clip should be engaged to prevent a stuck clutch—which can occur if the empty space between the clutch cones isn't maintained when the tractor sits unused for long periods. Use the U-clip to keep your clutch lever up, so the clutch is disengaged by leaving a space between the two clutch plates.*

When the OPC lever (red lever in image) is released, it either shuts off the engine (mechanical clutch models) or stops the hydraulic pressure, which disconnects the clutch plates and stops power transfer (hydro-mechanical models). In both cases, the tractor stops! With mechanical models, the OPC lever includes a U-shaped clip wire to keep the presence control down for starting and for idling in the field. PowerSafe models don't need this U-clip feature because the engine stays on when the OPC lever is released.

EQUIPMENT FEATURE

FULLY ADJUSTABLE HANDLEBARS

Handlebars on M2w-tractors can be adjusted: up and down for different operator heights, side to side (both left and right side) for better control, and fully 180° for different implement use (see Figure 2).

HORIZONTAL (OFFSETTING) ADJUSTMENT BENEFITS

- Avoids compaction of raised bed by the operator, who can walk in the path instead of on the bed top.
- Provides operator some elbow room when operating near obstacles, like a fence.
- Gives better visibility when snow-throwing to the left and walking offset on the right.
- Is more ergonomic and gives better control when mowing across a slope because the operator remains vertical.
- Offers a horizontal 180° turn to switch between front-mount mode and rear-mount mode for implements.

VERTICAL ADJUSTMENT BENEFITS

- Ability to adjust to operator's height improves ergonomics.
- Can be set for best operator position for many different tasks and conditions, like standing slightly above the implement on the unbroken ground to one side when trenching.
- Adjustable for storage or transport simply by folding the handlebars completely down.
- A quick adjustment of the handlebars down at the end of a row allows operator to lift a heavy soil-working implement out of the ground (such as a plow, tiller, or power harrow).

Pro Tip: *The old adage "lift with your knees, not with your back" applies when moving implements out of the soil: lower the handlebars, bend your knees, and then lift upward by straightening your knees.*

TRANSMISSION

Engine power transfers through the transmission to the axle, wheels, and the PTO to power implements. Tractor performance for different tasks is affected by transmission options: all-gear vs hydrostatic drive, with or without differential gearing, and having a mechanical or a hydro-mechanical clutch. Row crop tractors usually have a chain/belt drive.

DESIGN BOX

WHICH GEAR FOR THE JOB?

- **Seedbed preparation:** 1st gear for rear-tine tillers or power harrows. Going too fast creates an uneven seedbed, so stay in 1st to avoid additional corrective passes.
- **Plowing:** 1st gear in never-plowed earth, stony ground, and sloping sites and 2nd gear for seasonal plowing of already-prepared ground and when soil and topography allow.
- **Mowing:** 1st gear for initial mowing tasks with rotary-type mowers (lawn, brush, and flail) and thicker/taller material. 2nd gear for finish mowing. Sickle bar mowers can be operated in 2nd gear for most work, such as cutting hay; shifting down may be helpful when working around and under picket fences. 3rd gear is possible only with the dual-action sickle bar mower because the mowing action is much smoother. ***Note:*** Higher speeds are best in gentler terrain and when the grass is dry and not too thick.
- **Row cultivation:** 1st and 2nd gear. Faster speeds help cultivation, especially for straightforward single-row hilling. Too fast, and the inexperienced operator will harm crops in the row.
- **Plot cultivation:** 2nd gear for larger plot cultivation with S-tines to remove grass. Speed helps drag grass roots to the surface to dry out and die.
- **Bed forming:** 1st gear for *initial* path furrowing or bed forming using power ridger or rotary plows. 2nd gear for reforming with hiller/furrower or power ridger.

WHY CHOOSE A CERTAIN TRANSMISSION?

1. ALL-GEAR DRIVE

An **all-gear drive** is a mechanical drive system that runs *both the PTO and the wheels.* Similar to manual cars, you shift between neutral and different gear speeds. Most models have 2 or 3 speeds; others (like the BCS-853) have a 4th road speed. On *standard* tractors, engagement of the gears and clutch is carried out by levers, cables, and rods.

Benefits: The ease of selecting a known and constant operating speed. This is especially useful for long runs in a garden plot or when traveling on laneways. In general, all-gear drive systems are simpler to repair, so they may be a better option for DIY-types who live in more remote areas. ***Safety:*** Some operators find a mechanical drive system safer because *accidental* speed change is less likely to occur than with a hydrostatic drive. An all-gear drive is great for market growers and row crop farmers because a single, constant gear is mostly used for soil working.

2. HYDROSTATIC DRIVE

This transmission uses an all-gear system for the PTO, *but the wheels use a* **hydrostatic drive**, similar to an automatic car that uses hydraulic fluid to transmit the energy needed. This means the tractor can go forward or in reverse *without disengaging the PTO.* This design provides infinitely variable speed with a convenient handlebar lever. Operators can push the lever forward incrementally from zero to full rather than selecting set speeds 1, 2, 3, or 4.

POWERSAFE CLUTCH AND TRANSMISSION

For tractor nerds, here is the standard PowerSafe all-gear drive system that maximizes the energy output for high efficiency to the wheels and PTO shaft. A) Oil Pump Drive, B) Clutch, C) Transmission Drain Plug, D1) Operator Presence Control (OPC) switch, D2) Clutch Lever, E) Input Shaft, F) Worm Gear, G) Crown Gear (on larger models, this is the Differential Gear), H) Wheel Axle, I) Transmission Case (oil bath), J) Speed Gears, K) Reverse Gear, L) Gear Selector Rod, M) PTO Connection Point, N) Gear Selector Options Panel, O) Transmission Dipstick

IMAGE CREDIT: BCS AMERICA, INFOGRAPHIC CREDIT: ZACH LOEKS

Benefits: Speed control is more nuanced, which is preferable when driving distances within properties and for mowing tasks. Landscapers, grassland farmers, and orchard growers will benefit from this drive system. Also, the ability to change speed and direction without stopping the "live PTO" power transfer to implements is very helpful for back-and-forth fine mowing.

TYPES OF CLUTCHES AND OPERATOR PRESENCE CONTROLS

Mechanical or **hydro-mechanical** clutches allow the operator to connect and disconnect the flow of energy from the engine to the transmission without needing to stop the engine.

MECHANICAL (DRY) CLUTCHES

Mechanical clutches include a double-cone, spring-loaded clutch. Of import here, is that this is a *dry* clutch; it is located outside of the engine block, so it's a simpler system than wet clutch designs (see below). Squeezing the clutch lever upward activates the cable and engages the spring-loaded clutch, separating two cones about a ¼". Releasing the lever presses the two cones together to give a positive transfer of power from the engine to the gearbox and the wheels at the speed of the newly selected gear.

Why choose a mechanical clutch? They are a tried-and-true, efficient, and DIY-repair-friendly clutching system with an OPC that shuts off the engine when the lever is released.

HYDRO-MECHANICAL (WET) CLUTCH

The more recently invented hydro-mechanical (wet) clutch has similar efficiency, but instead of a spring action, *hydraulic pressure* is used for clutching. Here two cables (clutch and OPC cable) are actually connected to hydraulic valves. ***Note:*** Hydro-mechanical clutches are options with both transmissions: all-gear or hydrostatic.

Why choose a hydro-mechanical clutch? The OPC "dump valve" releases hydraulic pressure and creates an instant stop without shutting off the engine. This may be safer because an implement will stop faster when this type of OPC is released. This also means restarting the engine isn't necessary if the lever is released, which is helpful for pausing in the field. Hydraulic clutches run cooler, which is great for powerful implements (like a flail mower), especially if it is working hard and generates heat in the clutch. This can be more problematic with a dry clutch, especially when working hard for extended periods.

Pro Tip: *Tractors jump when hitting rocks, causing OPC release and stopping the tractor. This commonly occurs when incorrectly pressing down on the handlebars to force the tiller deeper. Doing this shifts the weight from the tires to the tiller's tines, causing the tines to actually drive the tractor forward (called tine-walking). Instead, tiller depth should be set by adjusting the implement and working in proper soil conditions.*

SIDE BOX

Hydro-Mechanical Clutch Maintenance Considerations

- Hydro-mechanical clutches use the transmission's fluid for hydraulic pressure and gear lubrication, so you need to use a **universal tractor fluid** (instead of gear oil), which requires more frequent changing since it serves more functions.
- An additional oil filter needs changing every 100 hours.
- Fully hydrostatic tractors also have a cooling radiator, which needs to be kept clean for optimal performance.

DIFFERENTIAL DRIVE/LOCKOUT AND BRAKES

The 853's differential drive sends power to the left and right wheels independently, allowing on-a-dime turning. With differential drive, when one wheel starts to spin more, the other wheel stops turning. **Locking the differential** puts full power to both wheels for increased traction. Differential drive is great for growers because it works in many different situations, conditions, and topographies.

Pro Tip: *It takes ¼ turn of the wheel for the differential to engage and switch between locked differential for field plowing and unlocked for turning at headlands. So, when locking the differential to get going, turn the wheels a bit by shifting the handlebars from side to side and disengage the differential 3–4 feet before it's needed again at the end of the row.*

WHEN DO I LOCK THE DIFFERENTIAL?

- When making long runs with earthworking implements, rotary plowing a 100 ft plot, or mowing a hay field.
- When operating in bad soil conditions that prompt the need for extra traction.
- When driving up a steep hill and you need to put power equally into both wheels.
- When the tractor is stuck, so you need both wheels working to pull it out of the hole or snow-filled ditch!
- When pulling a heavy load in a trailer along a straight lane.

WHEN DO I UNLOCK THE DIFFERENTIAL?

- When at the end of the row. Unlocking the differential makes for easy turning.
- When mowing and making many turns around a lawn area.
- When steering the tractor into a storage space.
- When operating the tractor between fields.
- When negotiating trails in a forest.

BACK-END

This is the tractor's business-end. It's where the PTO powerfully connects to implements and the clevis hitch attaches other equipment. Of course, the back-end can become the front when the handlebars are reversed.

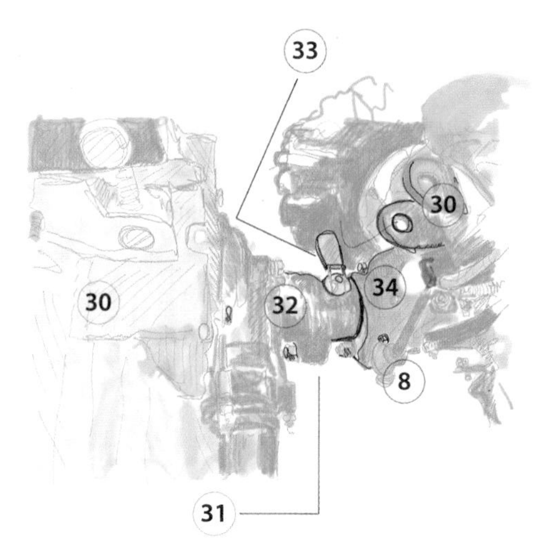

30. ***Implement***
31. ***Quick-Hitch System (male tang and female bushing)***
32. ***Quick-Hitch Tang (male) bolted to attachment***
33. ***Quick-Hitch Bushing (female) with Locking Lever***
34. ***Tractor—Rear Transmission Cover and PTO location***
8. ***PTO Engagement Rod***

AXLES AND WHEELS

This is where the "two-wheel" tractor gets its name—from its unique axle/wheel configuration. The axle transfers power from the gear box to the wheels when the clutch is released to engage the selected gear/speed. There are really two types of axles: straight axles (such as on the BCS 722), and models with differential drive (like the BCS 853), which actually have two half-axles. Some models with differential also include left/right steering brakes (located on the left and right handlebars).

TRACTOR WEIGHT AND ADDITIONAL WEIGHTS

The BCS 853 and 30" tiller together weigh 342 pounds. This is less than a riding mower, which means less compaction in the garden. However, the two-wheel tractor is *meant* to have additional weight for certain applications, and this is added to the front bumper, wheels, and directly onto implements. **Bumper weights** help balance the front-end when using heavy equipment (such as a power harrow). **Wheel weights** improve traction for steep slopes and sticky soils and are added to the wheel directly or through a set of barbell hangers. Wheel weights are essential for operating implements like the moldboard plow, subsoiler, and plastic mulch layer. **Implement weights** help the equipment dig in deeper to do a better job, or they can act as counterweight for the engine to keep lighter implements (like the narrower sickle bar mowers or snow throwers) on the ground.

Pro Tip: *The two-wheel tractor teeter-totter of weight: front- and back-end are balanced over the axle. If the front-end is too heavy, it will dig into the soil, causing the operator to force the tiller into the ground. Conversely, a too-heavy back-end results in traction loss at the wheels. Tractor models, weights, and implements must be selected for compatibility.*

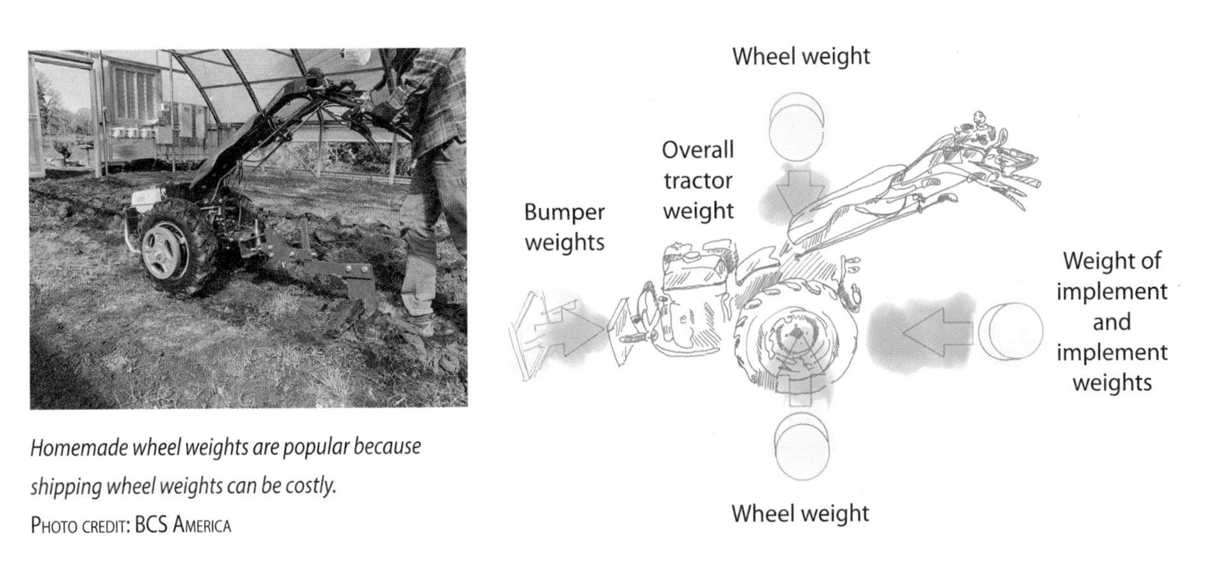

Homemade wheel weights are popular because shipping wheel weights can be costly.
Photo credit: BCS America

WHEEL TRACK WIDTH AND EXTENSION

The **wheel track** is best understood as three key measurements that affect garden design, tractor stability, crop cultivation, and more. Adjusting the wheel track width changes tractor performance for specific uses. We will discuss adjustments to wheel track width in Chapter 6.

WHEEL TRACK MEASUREMENTS

- **Inside-to-inside** measurement of wheels. For row cropping, this space determines crop row and cultivator placement, and adjustment allows passing over more or fewer crop rows and/or entire garden beds.
- **Center-to-center** measurement of wheels. For some growers, the wheel centers line up with garden path centers, and the tractor straddles the garden bed without driving on the growing area. Regardless of whether you straddle a garden bed or drive on the bed top, wheel track cultivators are centered behind wheels to erase tire marks.

- **Outside-to-outside** measurement of wheels is the widest point and determines *adjustments of path and row width*. If wanting to drive in paths (or between rows), say for reforming beds with a power ridger, then this measurement cannot be greater than path width.
- ***Note:*** Wheel track width is also affected by wheel type, width, and the use of double wheels.

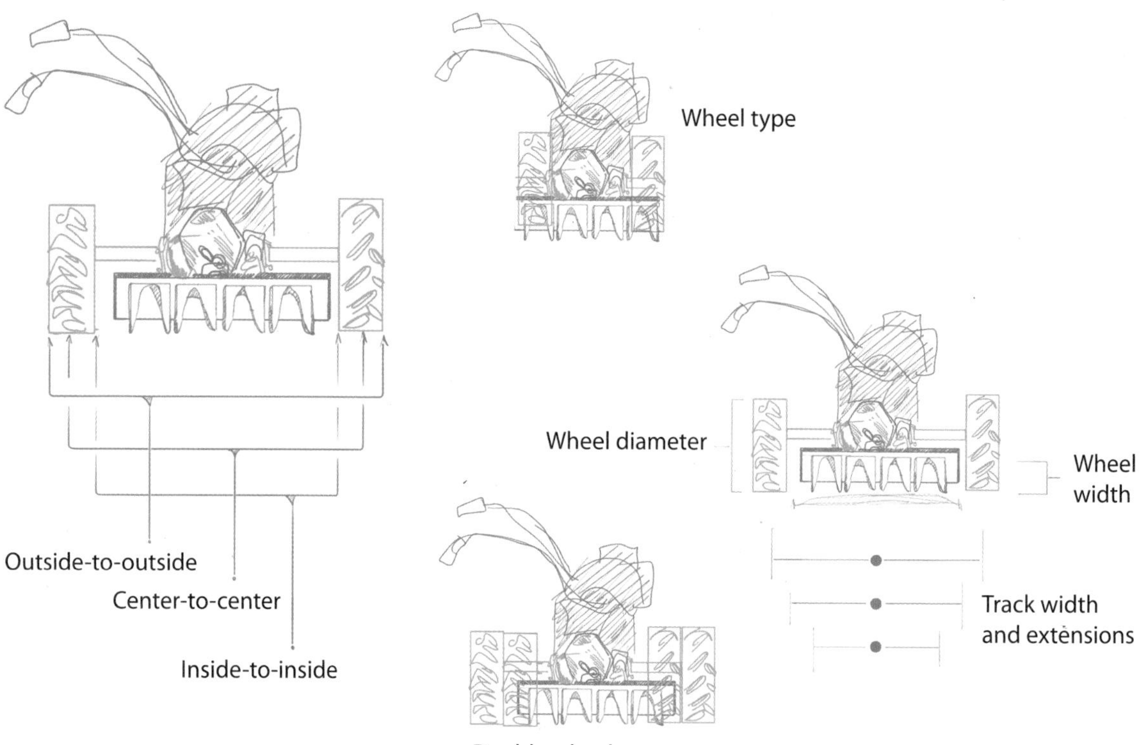

WHEELS AND TIRE SIZE OPTIONS

Different wheel types and tire sizes exist for unique applications.

- **Increasing wheel size (diameter):** This improves crop clearance, facilitates driving up onto raised bed tops and negotiating bumpy terrain, and increases ground speed. Certain implements may require a different wheel size for a particular tractor model. **Example:** The 5" × 12" × 22" wheel is standard on the BCS 853, which works great for the rotary plow. Conversely, the BCS 732 *requires* a 5" × 10" × 19" wheel instead of the standard 4" × 10" × 18" for improved plow operation.

- **Tire type:** There are many tire types and options. Which tires you use will alter tractor compaction, traction, and stability. Turf tires reduce compaction for lawn management; spike wheels help with cross-hillside mowing; cage wheels in tandem help with traction, and so do chains for winter use.

TRACTOR OPERATION: GETTING STARTED

This section helps get growers going! It provides some essential tips and tricks for the proper starting, operation, and safety of two-wheel tractors.

CHECK IT BEFORE STARTING

Before operating your tractor, there are some important items to check.

1. Tractor should have sufficient engine and transmission oil and fuel. Check this.
2. Keep implements clean and free of obstacles, such as twine wrapped around tiller tines.
3. Lubricate PTO shaft and grease implement nipples.

STEPS FOR TRACTOR STARTING

You can refer back to the Tractor Component Index (see figures: 3,4,5) for reference when following these steps.

1. Switch on the engine on/off switch (#15) for applicable models.
2. Move the throttle lever (#19) to halfway to regulate the fuel/air mix that enters the engine.
3. Put wheel speed selector rod (#9) (also referred to as the gear selector rod) to neutral (standard models); the hydrostatic speed control lever (#20) is set to neutral for hydrostatic models.
4. Keep shuttle reverse lever (#18), for standard models, in the forward setting.

5. Make sure the fuel valve is open using the fuel on/off lever (#2b).
6. Put the choke lever (#2a) to the left (closed position).
7. Pull gently on the recoil rope (#1) until your feel tension, then pull firm and straight, and release to start the engine. Turn the switch (10a) for electric start models or use the optional recoil rope that came with the tractor.
8. Now, open up the choke (#2a) to allow engine to run normally. If you leave the choke on too long, the engine will run bumpy and smoke will come out of the exhaust because there is too much fuel after the engine is already warm. With the choke open to the correct amount, you will hear and feel the engine run smoothly.

Pro Tip: *Pulling the recoil cable engages the crankshaft and turns the flywheel, which fires the spark plug and starts the piston strokes, which keeps the flywheel running and the crankshaft turning. The electric start uses energy from the battery (rather than human power) to charge a motor that spins the crankshaft.*

EQUIPMENT FEATURE

COLD STARTS

The choke can be closed for starting, especially for cold starts. By closing the choke, you are literally "choking off" air supply entering the carburetor where it mixes with fuel injected into the engine cylinder. Air is necessary for combustion, but when your engine is cold, it cannot burn small amounts of fuel, so it needs more. When your choke is closed with less air being pushed in, there is more pressure to pull in more fuel into the combustion chamber.

FIGURE 7: GETTING READY TO WORK WITH A BCS 770

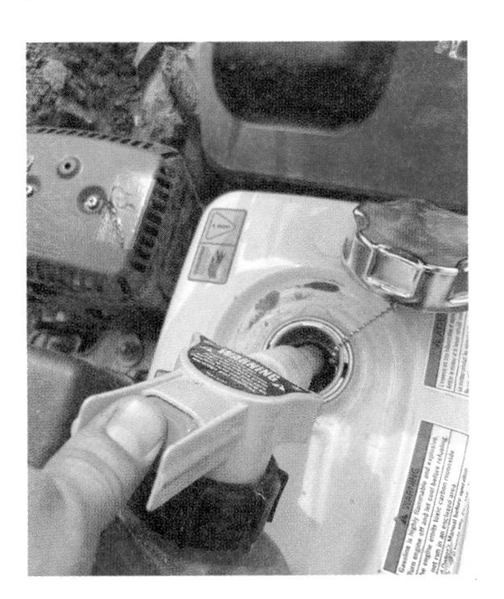

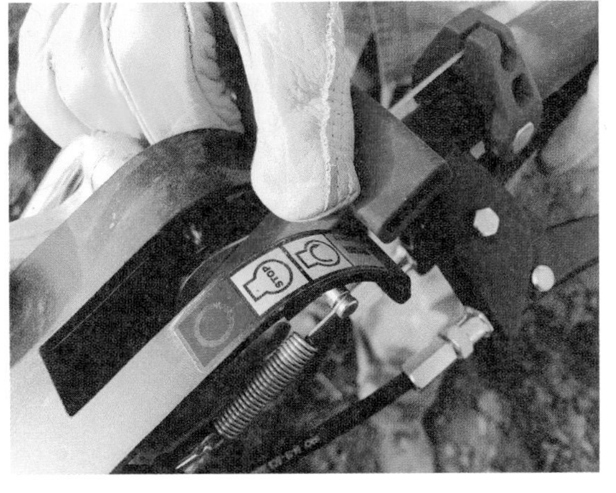

Top left clockwise: *Filling up with a spill-proof jerrican, engine on/off switch, hydrostatic speed selection set to neutral, hand throttling down, recoil start, differential lock.*

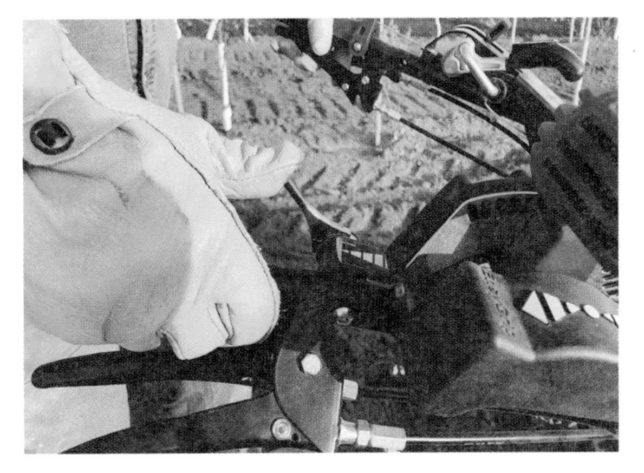

STEPS TO GET OPERATING

1. Put tractor in rear- or front-mount mode. For front-mount mode, reverse is selected *to move forward!*
2. Set handlebar height and offset according to the job.
3. Squeeze the clutch, then lower the OPC lever (on PowerSafe models), or lower the OPC before the clutch lever (on non-PowerSafe models).
4. **Standard transmission models:** Engage clutch, then select operating gear. **Hydrostatic models:** Select a high or low range with the selector rod.
5. Keep hand throttle at part-way open for starting.
6. **Standard models:** Release clutch and move forward. **Hydrostatic models:** Move the hydrostatic control lever forward gently. ***CAUTION:*** This lever is very responsive; if you push it all the way forward, the tractor will go at top speed. However, if the tractor does bound forward, you will be forced to let go of the OPC lever, stopping the power to the wheels.

Pro Tip: *If you slowly work up to a higher speed, your body will also be in the right gear to move forward with the tractor safely.*

STEPS TO BEGIN IMPLEMENT OPERATION

1. Situate tractor correctly for specific tasks before engaging PTO or releasing the clutch to move forward with any implement, this avoids issues such as unwanted tilling of lawn or irrigation lines.
2. Adjust implements ahead of time. For instance, lower roller cage on power harrow or set depth on a rear-tine tiller.
3. Lock/unlock differential as needed. Remember it takes forward motion for the differential to actually engage/disengage.
4. **For PTO-powered implements:** Squeeze the clutch lever and engage the PTO control rod.
5. Release clutch lever, and the PTO-implement will begin to operate with the tractor moving ahead OR the movement forward will begin to allow the draft or ground-driven implement to work.
6. Watch implement during operation for proper functioning. Stop and adjust settings as needed.

Pro Tip: *For implements to perform correctly from the get-go, set them up right from the start. However, it takes some operation time in the field and minor adjustments to get anything working the way you like it. Make notes of the best adjustments!*

A BIT OF TRACTOR SAFETY

Yes, protect yourself against moving sickle bar blades! However, some of the most pervasive dangers *are less obvious.* Those tiny daily injuries caused by repetitive motion, noise, and poor ergonomics can add up to lifelong harm. Safety is about: **design, maintenance, operation,** and **protective gear.**

Pro Tip: *Actually read your manual, observe safety stickers, and gain experience. Seriously.*

SAFE DESIGN

New two-wheel tractors are designed for ergonomics and safety. If you are planning on long hours of operation, these safety features are a blessing. As an edible ecosystem installer, I can spend 8 hours on a tractor some days!

DESIGNED SAFETY

- **Longer handlebars,** available with differential drive models, allow more operator room and leverage.
- **Vibration reduction and distribution** avoid passing it to the operator. **Examples:** low-vibration engines and shock-mounted steering columns.
- **Adjustable handlebar height and angle** are critical for operator comfort and proper positioning.
- **Front- and rear-mount modes** revolutionized the safer operation of certain implements.
- **Task-specific design** means you should use tractors and implements for their intended jobs. This doesn't mean you shouldn't innovate uses for equipment! The power ridger was designed for trenching to plant sweet potatoes, but it turns out that it is also great for building raised beds and furrowing to plant fruit trees. But that doesn't mean you should grind a stump with it … get a stump grinder!

Pro Tip: *We ask a lot of our equipment, and we ask too much of it when we don't provide routine maintenance! A seasonal maintenance schedule is key.*

SAFE OPERATION

Safe operation requires attention to equipment limits, operation conditions, and proper techniques.

SOME OPERATION SAFETY TIPS

- ☐ Remember, some tractors and implements appear similar but aren't—understand the differences.
- ☐ Scout surroundings for hazards and remove any obstacles like old twine, wire, stones, etc.
- ☐ Don't run a tractor in a closed space (except to move it out); tractor emissions are toxic.
- ☐ Never tie down the OPC when operating moving equipment. Keep your hand on it.
- ☐ Never lift unattached implements without proper strength and posture; lifting often requires two people.
- ☐ Don't put feet under the implements or touch implements with your hand or body when they are operating.
- ☐ Only operate on the opposite side of soil discharge for plow-type implements.
- ☐ Don't push the tractor with your body; allow the tractor's own weight and power to do the work. If you feel you are pushing, it is likely the soil conditions aren't correct (too wet) or the implement is incorrectly adjusted.
- ☐ Operate according to specific conditions and implements. **Example:** Stay downhill of the tractor and mower when operating across a slope. **Example:** Keep left of the rotary plow and tractor with offset handlebars so it doesn't spit soil and rocks *toward* you.
- ☐ Operate with a good footing on stable ground.

PROTECTIVE GEAR

Always wear protection! Unfortunately, this is a rule many people ignore. Awareness of small repetitive injuries is crucial if you are to avoid long-term injury. I got this advice from my wife: "Darling, wear your earmuffs, I want you to be able to hear my voice later in life." Then she put her own on and kept hilling the potatoes. Safety gear kits can be assembled for each operator.

ESSENTIAL PROTECTIVE GEAR

- **Noise-Dampening Ear Protection** should be worn for all operational uses.
- **Anti-Vibration Work Gloves** are essential since your hands are literally handling the controls and steering.
- **Steel-Toed Boots** keep your toes safe if you accidentally step near an active implement's tines or blades.
- **Safety Glasses** are recommended when using chippers and implements that can release excessive debris or dust.
- **Elbow Braces** are less known, but they are highly effective for dealing with the vibration that moves up the arm.
- **Wrist Braces** are helpful if extra support is desired.
- **Fitted Clothing** is needed. This is especially true when using implements like a chipper/shredder where the operator is directly engaged with the implement, feeding in wood. Save baggy fashion for dance night!
- **Protective Clothing** with padded knees is advisable for doing maintenance and taking implements on/off.
- **Work Hats** keep the sun off the head, reducing field fatigue and maintaining focus. Don't lose your cool!
- **Back Braces** are recommended for those who need/want extra support.

CHAPTER 3
Implements and Their Uses

There are many implements available for professional and DIY land management—for every season. Let's look at different types of equipment and their suitability for certain tasks and their nuanced uses for different enterprises. Then in the next chapter, we will dive into how to avoid over- or under-investing in equipment and how to design for scaling-up.

TYPES OF IMPLEMENTS

Let's look at the two main types of implements and their variations. ***Note:*** The different terminology for implement types used here is meant for a general overview; when in the field, growers call a plow *a plow*, and of course, they don't refer to a *non-PTO pull-type implement*; but the distinction helps when trying to understand how implements work and how you can plan their use for different field operations.

NON-PTO PULL-TYPE IMPLEMENTS

DRAFT IMPLEMENTS

Draft implements are pull-type and require no PTO power. Examples of draft implements are furrowers, root diggers, and subsoilers. They are the original farm equipment and are, of course, still relevant today. The initial purchase cost for draft implements is much lower than for PTO-powered implements. Some implements have a draft *and* a PTO version; a moldboard plow is a draft implement, but a rotary plow is a PTO-driven implement best used with M2w-tractors.

CULTIVATION IMPLEMENTS

Cultivation implements include S-tines for pulling up grass roots, gangs of cultivators for weeding row crops, and discs for hilling potatoes, corn, and

Plow

Furrower Root Digger

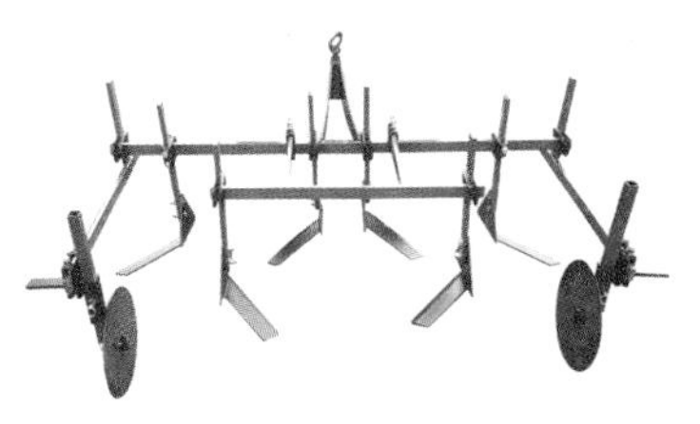

3-Row Tender Plant Hoe.
Photo credit: Thiessen Equipment

beans. Cultivators are often customizable to row systems and are returning in popularity—an homage to the way we used to farm. Best used with row crop tractors.

SIDE BOX

Depth and Implement Control

Many implements have **gauge wheel(s)** to control depth and keep them centered on rows. The rotary plow gauge wheel is adjusted up/down with a pin to set plow depth. Fully-extending gauge wheels is good for transport between jobs and at headlands when turning. Also, **cage rollers** on some implements, like power harrows and rear-tine tillers, set tine depth. These rollers have the dual purpose of pressing and firming the seedbed. There is also a **planar wedge** on the underside of the rear-tine tiller that acts like a shoe pressing into the soil to prevent the tines from going deeper.

GROUND-DRIVEN IMPLEMENTS

These pull-type non-PTO implements have active parts engaged by the ground itself or are driven by the wheels as the tractor pulls them forward. **Compost spreaders** are a good example. But cultivation equipment like **basket weeders** are also ground-driven. On the underside of the **finger**

weeder are metal spikes that engage the soil to drive the weeder's rubber fingers, allowing very effective in-row weeding of crop rows. The **3-row seeder** is a classic example, where the seeder's wheels drive the seed-dropping mechanism. Many of these implements are discussed later.

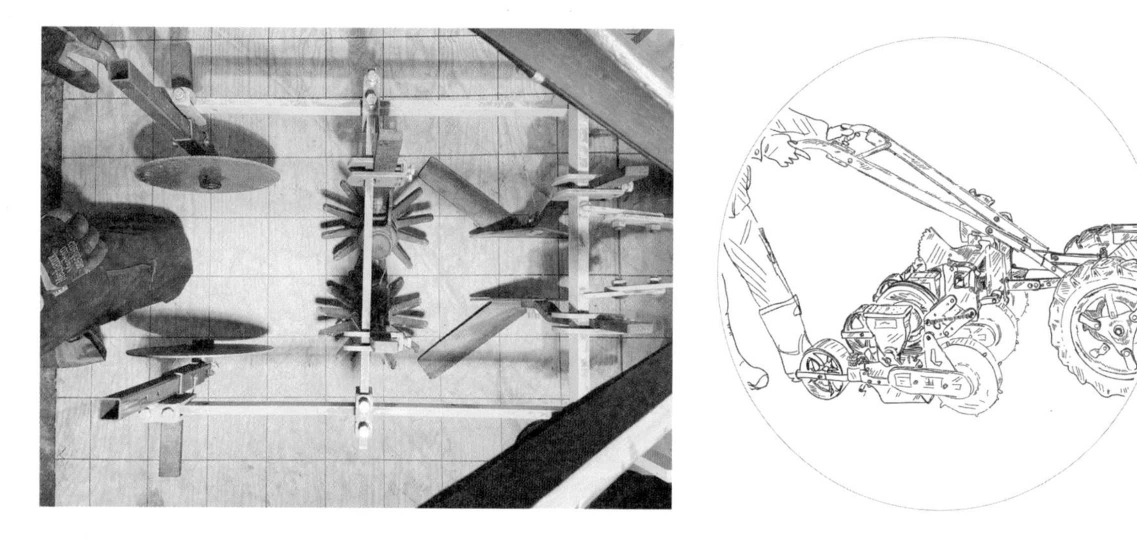

NON-PTO PUSH-TYPE IMPLEMENTS

There aren't many push-type implements that *aren't PTO-powered.* The snow blade and buddy cart are examples of non-PTO push implements. Put the tractor in front-mount mode when using any push-type implements.

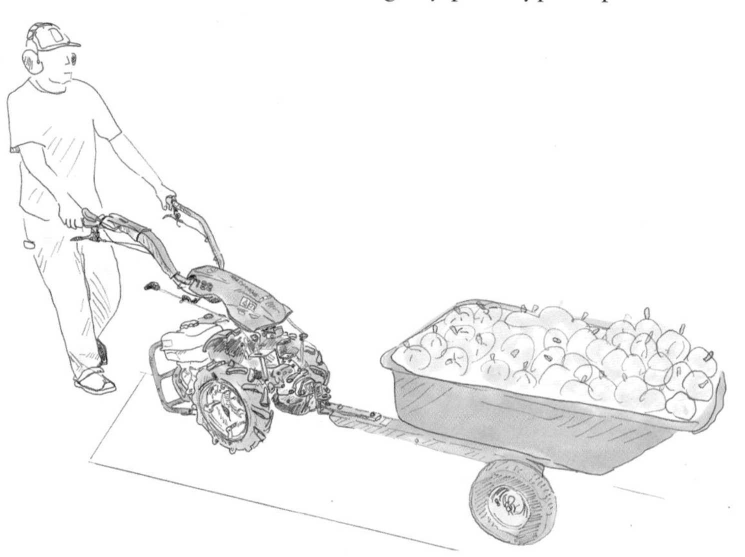

The buddy cart can be mounted to the tractor's PTO connection point with either a quick hitch or by bolting it directly to the PTO stud.

PTO-POWERED IMPLEMENTS

The most powerful types of implements require an M2w-tractor. Make sure you have the right PTO on your tractor for the implement; they are, unfortunately, not brand-universal; however, adaptors are available in many cases.

PTO-POWERED PULL-TYPE IMPLEMENT

PTO-powered *pull-type* implements are pulled by the tractor while it is in **rear-mount mode**. The active component is engaged by the PTO. Popular pull-type implements include tillers, power harrows, and rotary plows. The rotation of the handlebars (about 30°) allows the operator to walk alongside the tractor, which improves ergonomics and is safer.

The tiller on this Ferrari tractor is a PTO-powered pull-type implement.

Reid's sickle bar mower in Quebec is a PTO-powered push-type implement. On this tractor, the handlebars are rotated 180°, so the implement can be pushed in front of the tractor's wheels. Photo credit: Tourne-Sol Co-operative Farm

PTO-POWERED PUSH-TYPE IMPLEMENTS

The second type of PTO-powered implement is the *push-type*. These implements are pushed in front of the tractor while it is in **front-mount mode**, and the PTO engages the active components, such as the mowing knives on the sickle bar mower. Popular push-type implements include power ridgers and snow throwers, and sickle bar, lawn, and flail mowers.

GETTING IMPLEMENTS ON/OFF

Draft-type implements go on/off tractor hitches with a simple drop pin. Attachment of PTO implements requires more consideration. The proper installation of quick hitches makes it all easier.

Top left, clockwise: Putting implements on using the front-mount mode; Quick hitch and PTO on tractor; Implement with integrated tang, Close-up of dog teeth on implement's tang.

BASIC DIRECTIONS FOR CONNECTING IMPLEMENTS:

- Turn the engine off (preferably), put gear/speed selector in neutral, and disengage PTO.
- **Option 1:** Use front-mount mode and drive and/or roll tractor *toward* implement. This option provides good tractor control and no tripping-up on implements while facilitating the use of *blocking* (or walls) to help hold implements in place for connection. However, there is less visibility.
- **Option 2:** Use rear-mount mode with offset handlebars and drive tractor *backward* to implement. This option provides great visibility and allows

easy adjustment of implement without going around the tractor. However, this position can be tricky when attaching large implements (such as a mulch layer, chipper/shredder, or snow thrower).

- Keep the tractor throttle low and get the PTO right up to the implement before adjusting for connection.
- It is necessary for the dog teeth on the tractor's PTO *and* those on the implement's tang to mate *(remember: "hills into valleys")*. Rotate the shaft slightly until they mesh. ***Note:*** The PTO moves freely when disengaged.
- If using a quick hitch, lock the implement's tang into the female quick hitch bushing with the drop pin using the hitch lever. If no quick hitch is used, bolt the implement directly onto the PTO studs.

Pro Tip: *Dome side out! When bolting the quick hitch bushing to the tractor, or when bolting an attachment directly to the PTO studs (when not using a QH), it is EXTREMELY important that the PTO washers are installed with the dome side facing outwards. Failure to install the washers, or to install them facing the incorrect direction, can cause the PTO studs to become bent or even break off entirely.*

EQUIPMENT FEATURE

QUICK HITCH

The quick hitch is an essential accessory for serious operators. Here, a female bushing is attached to the tractor, and a male tang is connected to each implement. Disengage the lock pin with the lever to allow the implements to pull free and re-engage with another implement. Keep your PTO shaft well lubricated with grease and lubricate the lock pin as well with a lubricating spray.

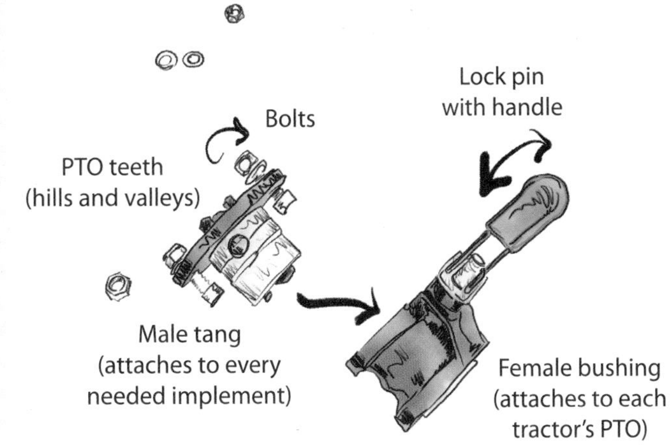

TIPS FOR IMPLEMENT CONNECTION:

- Keep everything greased, or the hitch system will seize up.
- Don't lift implements casually; most are heavy and should *not* be lifted by one person!
- Taking the implement off is straightforward. It is usually done just by rolling or driving the tractor back from the implement. A little shifting from side to side will slide it off.
- Keep implements on a level surface and have them positioned for easy hook-up.
- Plywood underneath and/or wood blocking behind reduces implement movement when connecting. This is particularly helpful with a power ridger.
- Cut pieces of 4" × 4" and 6" × 6" wooden posts for use as blocking behind and under implements.
- The rotary plow blades are tilted, resulting in *the two bolts* beside the implement's PTO tang to not line up with the receiving end on the tractor. Suggestion: drive back to the rotary plow until the plow's quick-hitch tang is inserted into the tractor's female bushing. With the tractor in park, the left hand on the plow handlebar and the right hand on the tractor's handlebar, lift the tractor's back-end by pushing up, carrying the rotary plow up with it. Now you can turn the plow so the two bolts line up and gravity takes care of the rest, with everything *falling* into place. Then, lock the pin.
- A sickle bar mower is easy to attach, but ensure that the *blade cover remains on* when attaching or removing the mower. That sickle bar is *sharp, exposed,* and *low-down* (out of operator view).

Pro Tip: *The pin can stick when conditions aren't level and the hitch isn't properly installed. Make sure the tang and bushing are clean for the best fit. If all else fails, a little tap with a small fine-tipped hammer helps lock it in.*

SELECTED EQUIPMENT PROFILES

I feature the following implements because they are the most universally applicable for key tasks. Many other implements exist and can be found as part of the further discussion about tractor users and operation cycles later in this book.

1. ROTARY PLOW AND SWIVEL ROTARY PLOW

USE AND APPLICATION

The rotary plow is highly multi-functional. At start-up, it is useful for plowing new ground to form a plot, specialized work like forming long mounds for bed building, and seasonal routines like plowing under cover crop. Additional functions include furrowing (for asparagus, leeks, or flower bulbs), hilling (for potatoes), burying (planted trees), and management of Compost-a-Path for in situ fertility. I will say more about these functions further on in the book.

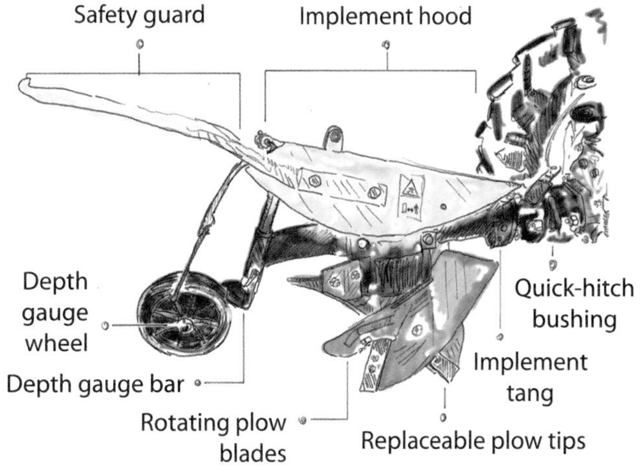

- **Safety:** Operate the rotary plow only when the handlebars are offset to the left and you are walking on the left to avoid soil from being launched at you.
- ***Note:*** If the rotary plow gets a buildup of clay or weedy debris, it will bounce along the soil surface instead of digging into the ground.
- **Alternative implement:** The power ridger is giving the rotary plow stiff competition for some of these functions, but it is not nearly as multi-functional and cannot plow a new field.

ACCESSORIES, OPTIONS, AND ADJUSTMENTS

The rotary plow is a PTO-powered pull-type implement used in rear-mount mode. Offset the handlebars and walk adjacent and to the left of the plow on unplowed ground. Removing the minor plow shield increases soil discharge for bed making and plowing; keep it on for bed reforming, hilling, and refined soil movement. Recommended inside-to-inside wheel track for rotary plows is ~15" to 18" to maximize the plow's *bite-full of soil,* opening more ground per pass.

- **Rotary Plow Accessories:** 5" × 10" wheels and a 5.5" wheel extension for some tractor models for rotary plowing.
- **Swivel Plow Accessories:** Bumper weight and 5.5" wheel extensions for some tractor models.

FARM FEATURE

FLOWER FARM USING ROTARY PLOW TO THE MAX

Gina at SHEGROWS operates a 1-acre flower farm in suburban Colorado. All the lawn was turned over by rotary plow and used to make raised beds that were finished using a rear-tine tiller. Many of her flowers are grown from bulbs, like dahlias. She used to run the hiller/furrower down the bed's centers for bulb planting. Her epiphany: "Why not plant the bulbs in the deeper paths made with the rotary plow?" Now, the rotary plow is used for bulb trenches. She grew her business to a static-scale in bursts of land expansion and equipment acquisition to grow her annual and perennial flowers to a 1-acre operation and help maintain a 3-acre property.

PHOTO CREDIT: GINA SCHLEY

EQUIPMENT FEATURE

THREE REASONS TO USE A SWIVEL ROTARY PLOW

The **swivel rotary plow** is a useful alternative for plowing slopes and larger plots because the blades swivel, allowing both *right and left soil discharge*. It is a big-time saver for specific situations. ***Safety:*** The swivel plow is heavy and needs extra strength to use. ***Note:*** A full description of plowing steps is in Chapter 7.

1. ***Big Fields:*** Normally, plowing is outward-moving and rectangular-concentric around the plot mid-line; this increases the distance between unplowed ground, requiring more **headland** travel. The swivel plow forms a furrow and returns using the same trench, reducing headland travel because *plowing from one side of the plot to the other is possible.*
2. **Sloped Plots:** When plowing sloped land, start at the top and work down, *not in the middle.* This is *not possible* with a normal plow, which needs to work from the center; a swivel plow *can* plow back and forth from the top down by swiveling the plow blade and using the same trench.
3. **Landscapers:** When building raised beds along fences, property lines, or laneways, landscapers need the swivel feature to plow in their chosen direction.

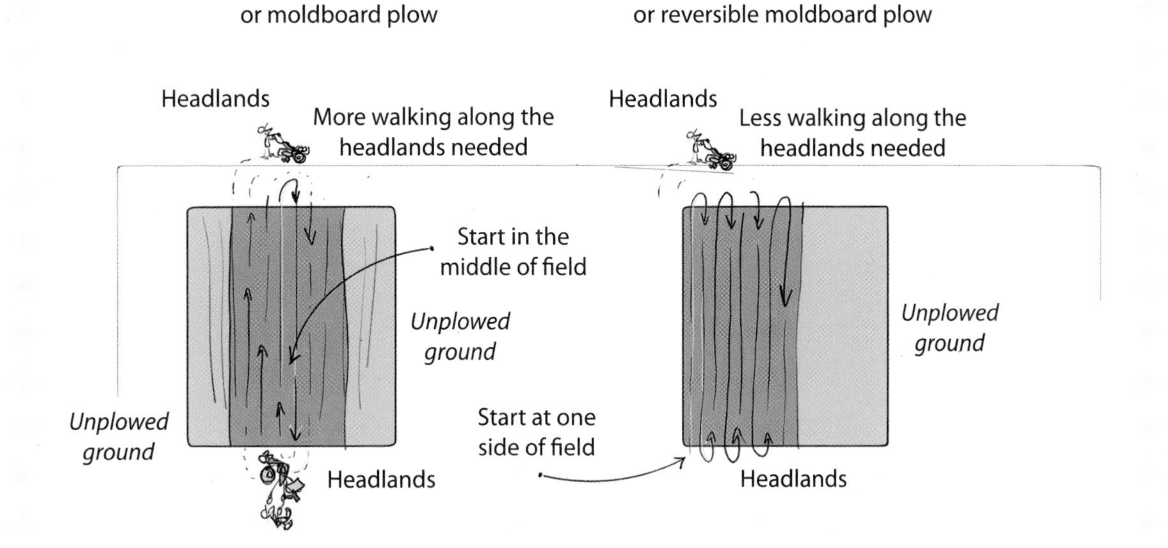

The well-designed rotary plow allows the brunt of the work to be taken up by replaceable blade tips!

2. PDR REAR-TINE TILLER

USE AND APPLICATION

The popular rear-tine tiller softens plow ridges, finishes seedbeds, and power composts green manures. Turning new ground *directly* into a garden (without any plowing at all) is only recommended for previously worked plots, not old fields. Otherwise, plow first, then pass with a tiller to remove ridges, leaving smooth, flat soil. The ***Precision Depth Roller (PDR)*** attaches to the tiller for depth control and presses the soil for better *soil-to-seed contact.* Tillers can flatten a *new* bed top (5–6" depth), re-prepare seedbeds (2–3" depth), turn cover crops under as green manure (3–5" depth) following a flail mower, and turn compost into the soil (2–3" depth) once it's been spread. Whenever you till, use S4-tillage principles, which are discussed in the Design Box, below, "S-4 Tillage Principles."

- **Safety:** Don't step too close to the tiller and never lift the tiller out of the soil while it is engaged.
- ***Note:*** Replacement tines are a good thing to have on hand to avoid slowing field work in stony ground.
- **Alternative implement:** Power harrow works for bed preparation, but not for cover crop management.

Pro Tip: *Consider marking "tiller killers" (aka large subsurface rocks) with flags so you can pull them out and avoid repeat incidents.*

DON'T TILL WEEDS, JUST PRE-WEED

We shouldn't rely on tillers to remove weeds. There are far easier, less costly, and more sustainable ways to weed. **Pre-weeding** (*stale seeding*) is the process of weeding the bed top before any crops have been seeded. Prepare the bed to seed, delay seeding for 5–9 days, and then make a pass with a tine weeder, rake, or flame weeder to kill young weeds at the *white thread stage* (when weeds are just barely visible). **Tarp culture**, in which black poly tarps cover the beds during a fallow period is also very effective.

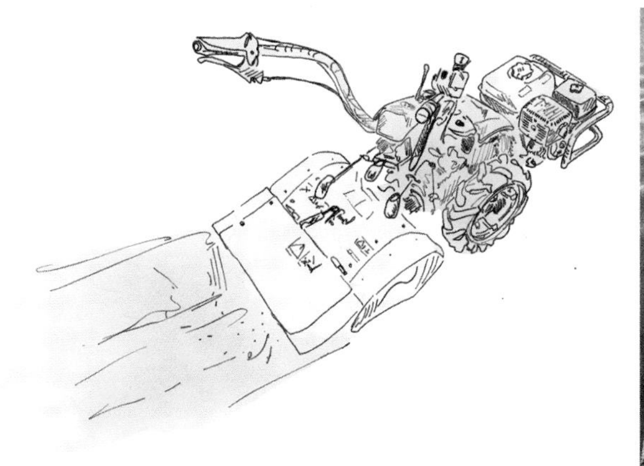

A rear-tine tiller without the PDR removes plow ridges, but the bed top is less firm and less even than when the PDR is properly adjusted.

ACCESSORIES, OPTIONS, AND ADJUSTMENTS

The PDR rear-tine tiller is a PTO-powered, pull-type implement used in rear-mount mode. Offset the handlebars and walk adjacent to the tiller. Adjust depth by raising/lowering the *planar wedge* and the PDR. The PDR has refined depth control (.5" increments), providing seedbed preparation similar to a *power harrow*—that quintessential bed-prep implement. A hiller/furrower can be attached to a rear-tine tiller to make *furrows* for potatoes or to make mini-ridges for sweet potato beds and other crops. Furrowers can make raised beds by *furrowing out* the paths between beds. Tiller widths range from 18" to 33". A standard 30" bed top can use a 30" tiller or two overlapping passes with a 26". More on bed forming techniques later.

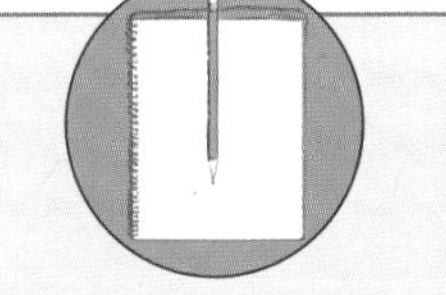

DESIGN BOX

STRIP TILLAGE

Tiller width is adjusted by removing tines for narrower tillage settings (from 33" to 27", 30" to 26", or from 26" to 20"). This can help with **strip tillage**, a soil conservation technique in which *only some growing space is tilled* to prepare for new crop seed/planting—leaving strips of original crop residue or cover crop as mulch.

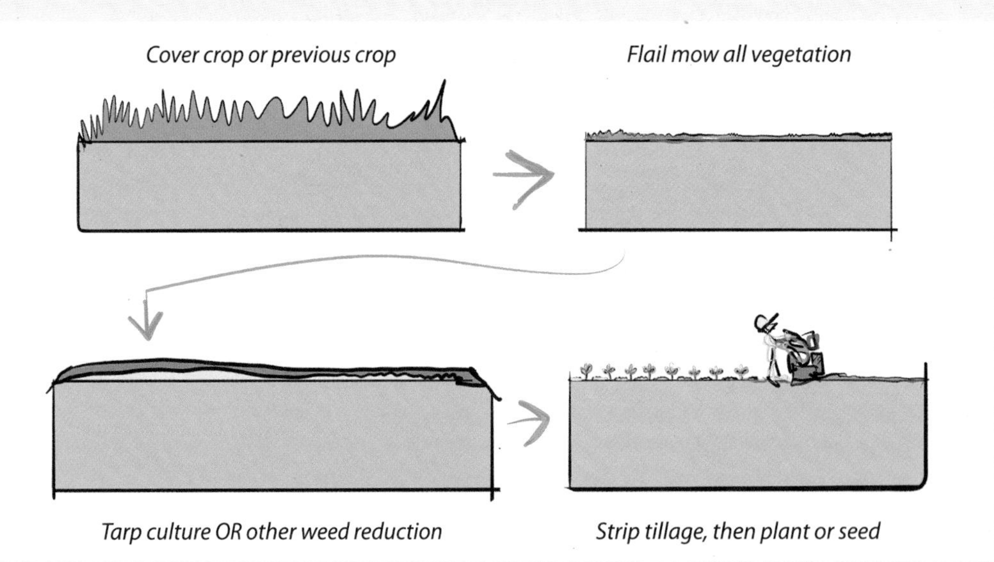

Pro Tip: *Strip tillage requires pre-weeding, so no hand weeding of the mulch is needed.*

Precision depth rollers allow growers access to all the typical multi-functional purposes of a rear-tine tiller, and they also offer excellent final seedbed preparation that improves seedbed firmness and consistency.

EARTHWORKS: TILLAGE AND FALLACY

Earthworks, or any earth moving (including tillage), is ubiquitous for many enterprises. It usually refers to initial landscaping (pond digging) and primary plowing. But really, all soil work can also be categorized under earthworks to reduce confusion about "tillage." Earthworks are essential to start-up for homesteaders, landscapers, and farmers using organic, Permaculture, or Ecosystem designs.

The term **tillage** usually refers to actual rototilling. Yet, originally it meant *working the soil*. Essentially, "tillage" goes back to early humans *dibbling* for wild seeds; it gained popularity with the advent of agriculture (~10,000 years ago), intensified with metal works, then industrial mechanization, and became insane at the end of the 20th century with 250 hp tractors that rip down fields. The extreme shouldn't frame small-scale earthworks; the science of soil is a better teacher!

It is important to define **no-till** *and* **low-tillage** as methods that *reduce* overall soil disturbance. But *in no way is it possible to eliminate* all earthworks from a farm! At the very least, *primary earthworking* is needed to open new land and/or build garden beds, and subsequent low-impact tilling techniques are highly beneficial to most operations.

Pro Tip: *Earthworks can be a great ally for soil conservation, disease management, and efficiency when using methods like Compost-a-Path—where heavy work like (cover) crop debris mowing and compost turning is done in-path and benefits are transferred to the bed top later, in the form of compost.*

FIGURE 8: COMMON PTO-POWERED EARTHWORKS

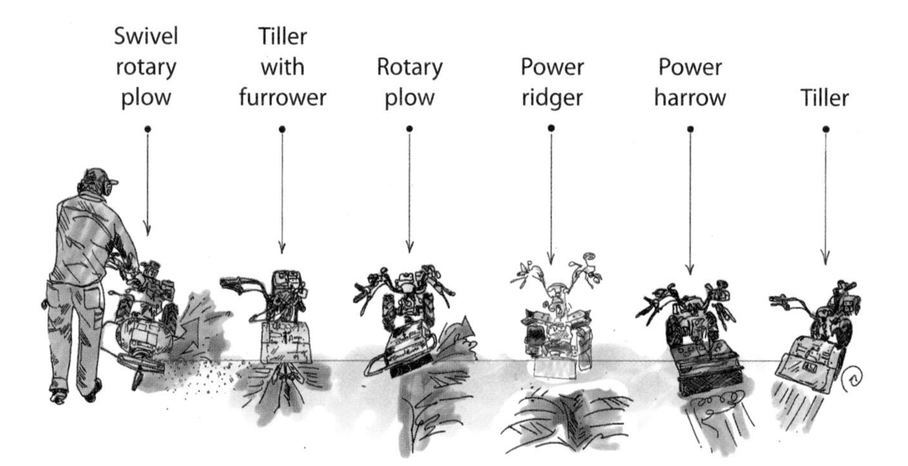

DESIGN BOX

S4-TILLAGE PRINCIPLES

Tillage rightly has a bad rap for 1) destroying soil aggregates, 2) creating compaction layers, 3) loss of nitrogen and nutrients, 4) loss of soil organic matter, 5) pulverizing of mycorrhizal fungi hyphae, and 6) the overall disturbance of soil ecosystem. A compaction layer (*plow pan*) results from excessive plowing at a constant depth and/or the tiller's slapping effect. Tillage benefits are achieved with lower impact using **S4-tillage principles:** *(Seldom, Shallow, Softly, Sorted).*

- **Seldom:** Only do essential tillage—in tandem with alternative techniques.
- **Shallow:** Adjust equipment for task-appropriate depth.
- **Softly (soft-tillage)**: *Soften the blow* of tillage through **soil horizon design** (see "Permabed

Soil Horizons" in Chapter 6), using task-specific horizons to buffer soil life communities in **aggregates** in an undisturbed **soil conservation core** (described in detail in the Permabed System chapter).

- **Sorted (patterned):** Organizing earthworks to avoid larger landscape issues of erosion from expanses of bare soil. A percentage of beds are always in cover or crop, and adjacent areas are never both tilled (alternate beds).

Pro Tip: *Permabeds and other raised garden beds are a key way to reduce soil compaction, preserve aggregates, and improve soil life communities. Consider higher Permabeds (4–6") in wetter climates for drainage and reduced compaction, and lower ones (2–4") for sandy soil to conserve water. For perennials, taller beds (6–8") improve root growth. I make 14" beds covered with weed barrier for husk cherries; when ripe, they roll right into the dry, clean aisle for easy picking.*

SORTED TILLAGE AS PART OF ALTERNATE LAND PATTERNING

Separate fields are commonly managed differently each year. Fields might be in production crops, cover crops, bare fallow, or tilled seedbeds. (A) **Alternate Land Patterning** subdivides plots into alternating *beds* (B) or alternating *cultivation zones* (C). The land is sorted into crops (b2) or cover crops (c1) and alternated with newly prepared beds (b1) or fallow fields (c2). Bare soil is susceptible to erosion and compaction. With alternate land patterning, adjacent land is not bare at the same time, reducing soil degradation without losing timely fieldwork.

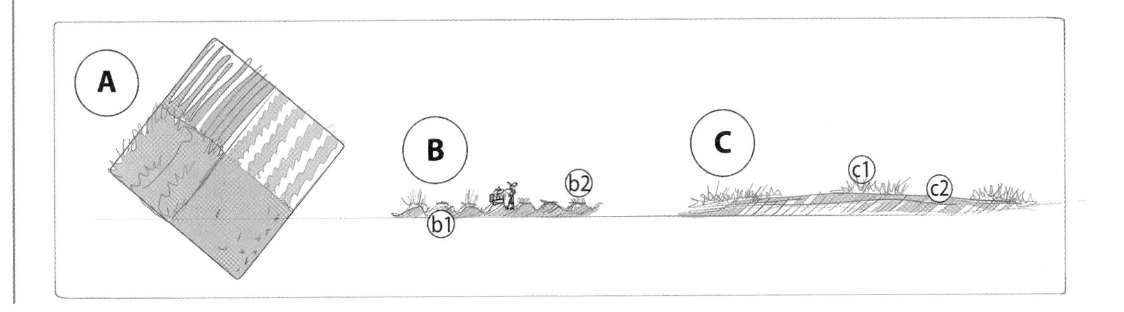

3. POWER RIDGER

USE AND APPLICATIONS

The power ridger is used for furrowing, trenching, hilling, and bed forming. Its *tiller-furrower-plowing* function loosens and jettisons soil out both sides (A). This leaves an open trench (easily planted), and ridges (could be new bed shoulders or hilling of crops). Apply slight upward pressure on handlebars to help implement dig in, then adjust handlebars to the correct height for walking behind *in the trench*. If continuous pressure is needed at the handlebars, then one of the following is true: a) soil conditions are too heavy, b) they are too wet, or c) not enough primary tillage has occurred.

Pro Tip: *A power ridger can be used for turning compost windrows using the Compost-a-Path Method (covered more fully, below) in conjunction with a 3-year debris management cycle—where heavy debris is followed by medium then light debris crops; then the paths are turned back onto the bed tops.*

ACCESSORIES, OPTIONS, AND ADJUSTMENTS

This is a PTO-powered push-type implement used in front-mount mode (B), set in reverse to go forward. The operator walks behind, in the trench. The depth of the trench can be adjusted by raising the depth gauge (C); this gauge also helps transport by lowering the wheel at plot-ends to travel headlands easily. Soil discharge is controlled by adjusting spring-loaded pins so the "wings" raise or lower (D).

4. FLAIL MOWER AND ROLLER ACCESSORY

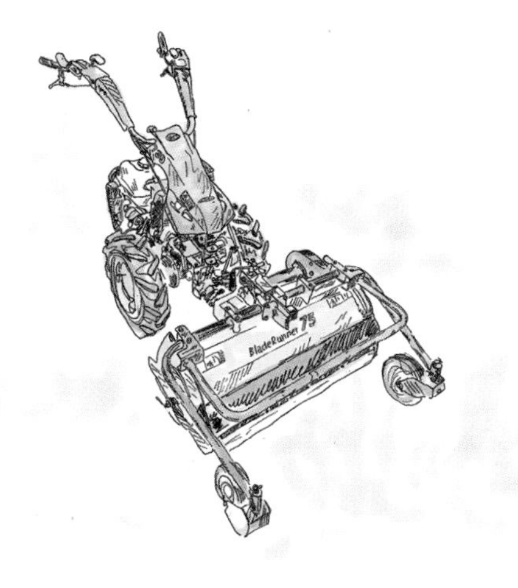

The flail mower is perhaps the most critical implement for sustainable market gardening and farming because it allows growers to integrate cover crops and make use of crop debris as in situ fertility, mulch, and compost.

USE AND APPLICATION

The quintessential flail mower chops crop debris and cover crops with knife-like blades to produce material that is coarser than what a lawnmower would produce, but finer than what a brush mower leaves behind. The ability to create well-chopped, but not super-fine, material is great for small-scale intensive growers. A flail mower makes quick work of thick-stalked kale, pepper, and eggplant debris, as well as tall green manures like winter rye, oats, and buckwheat. This equates to weed and fertility management and allows quick Permabed turnover for new vegetable successions and even in situ mulch management.

- **Safety:** Flail mowers can accidentally pull in wire, twine, and other debris that will tangle the flails.

- **Alternative implement:** Flail mowers are used to create in situ mulches: winter rye and/or vetch are mowed, and the debris is left on the surface of the soil to serve as mulch. A roller/crimper creates an even better weed-suppressing mulch, but timing using this technique is sensitive.

Pro Tip: *Growers will often flail mow crop debris or a cover crop, turn over the beds to further bury this debris, and allow 2–4 weeks of fallow time before using a rear-tine tiller to power compost the debris into the bed as green manure.*

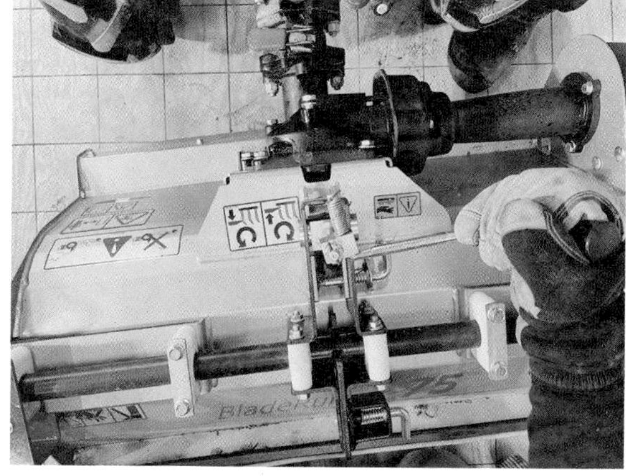

Adjusting the cut height and removing crop debris after a sweet potato harvest.

ACCESSORIES, OPTIONS, AND ADJUSTMENTS

This is a PTO-powered push-type implement used in front-mount mode set in reverse to go forward. Mowers can ride on their skids or on an optional roller; adjusting cut height is done simply by rotating the top-mounted handle to raise/lower the gauge wheels to create more/less clearance under the roller blades. There is also an L-pin used to flip the gauge wheels up for working through denser material, like a very tall stand of cover crop or weedy vegetation. There are grease nipples for the gauge wheels. Don't forget to grease them!

DESIGN BOX
FUMIGANT COVER CROPS

Bulb and stem nematodes (*Ditylenchus dipsaci*) decimate allium bulbs and roots, greatly reducing marketable garlic harvests. Insult to injury: fields remain infested for 10–20 years! However, when allied with a mustard cover crop, the flail mower and rear-tine tiller will work to fight garlic disease. Yes, of course, we must buy only disease-free seed, remove prematurely yellowing bulbs, and use clean equipment. But let's employ fumigant cover crops in addition to those best practices. Hot oriental mustard cover crops, like "Cutlass" and "Forge," are ecological fumigants that can reduce infestations and be used as a precautionary measure for new seed. **Method:** Seed a cover crop of hot mustards, flail mow when mature, immediately till into the soil, then water in (if it doesn't rain!). The result will be much lower nematode populations in the following year—results comparable to the use of commercial fumigants; however, unlike with synthetic fumigants, the populations may stay low in subsequent years, as well.

5. SICKLE BAR MOWER

USE AND APPLICATIONS

There are two main advantages of the sickle bar mower. First, this mower cuts and drops whole stalks to dry for hay, other forage, and even small grains production. Second, the mower has blades that extend well past the wheelbase, allowing cutting under obstacles. Its versatile T-shape can cut grassy edges along picket-style fences, moving in and out between posts, and maintain ground cover under orchard trees without the operator or tractor getting scratched or damaged. Also, this implement has an advantage around edges of ponds or ditches because the sickle bar can reach to mow along wet edges while the operator and tractor remain on firm ground.

- **Safety:** The sickle blade is sharp and out-of-sight under grass when cutting—pay attention, know the terrain (bumps, depressions), and don't cut near people!
- **Alternative equipment:** There is a single-purpose reaper-binder available in Europe (launching soon in America) that can do simultaneous cutting and binding of small grain. Although not an implement, it is a very useful single-purpose machine for scaling-up small-scale grain production.

Pro Tip: *Operate the sickle bar mower with the tractor throttle set at half open to limit the vibration. The dual-action sickle bar mower can be set at ⅔ throttle because its two-blade assemblies are moving opposite to one another, effectively reducing vibration.*

ADJUSTMENTS AND ACCESSORIES

This is a PTO-powered push-type implement used in front-mount mode set in reverse to go forward. Walk behind the mower. Adjust overlap of cutting passes about 3" to ensure a good, clean cut. Because these mowers are light, they benefit from a weight kit to keep the mower close to the ground. The adjustable skids set the ideal cut height, and its shaft easily follows ground contours.

- **Option:** A dual-action sickle bar mower is gaining popularity as a low-vibration model with wider cuts. The original sickle mower doesn't require larger tractors and is *affordable* for part-time operations. However, enterprises requiring a lot of mowing, like grassland farming, should opt for the dual-action mower to reduce operator fatigue and increase the cut rate when haying. ***Note:*** The dual-action mower has twice as many cuts per forward foot of motion, and the tractor can be run in 3rd gear—saving time for big jobs!

- **Maintenance tips:** The BCS sickle bar mower has an oil bath transmission which reduces maintenance. There is a grease nipple on the side of the transmission that lubricates the floating bar. Interestingly, it is the grass moisture that keeps the blade assembly lubricated, which is why it is important to only run the mower when you have it *in* the grass and are ready to start moving forward.

FARM FEATURE

SICKLE BAR MOWER FOR SPECIALTY SMALL FRUIT

Two-wheel tractors have a number of practical uses for specialty fruit production. Here, Sergio employs his two-wheel tractor to allow easy access to high tunnels full of highbush blueberries and other small fruit in the mountain valleys of Borgo Valsugana, Italy. He relies on the sickle bar mower to maintain the grass **leys** (mini fields that serve as access paths) between his berry shrubs. Greenhouse operations and perennial farms often find small-scale equipment solutions to meet their needs.

6. COMPOST SPREADER

USE AND APPLICATION

The spreader applies compost to garden beds or fields; the compost is then gently tilled in just before seeding new vegetable successions. A 1" spread is routine for seasonal vegetables, and 3" should be used for the initial establishment of new garden beds. Thick composting and cover cropping can get a soil engine running. Seasonal fertility applications should blend an organic compost base with nutrients according to soil test suggestions. Add kelp meal (nitrogen and potassium), ground crab/lobster shells (calcium and magnesium), and other essentials. **Example:** Neptune's Harvest fertilizer, which has a great range of nutrient blends. ***Safety:*** Don't adjust spreader thickness while the tractor is moving or put hands in compost to try to help it out of the opening. ***Note:*** The compost spreader should not be used for spreading manure or unfinished farm-made compost. Fine, granular material is necessary for an even spread.

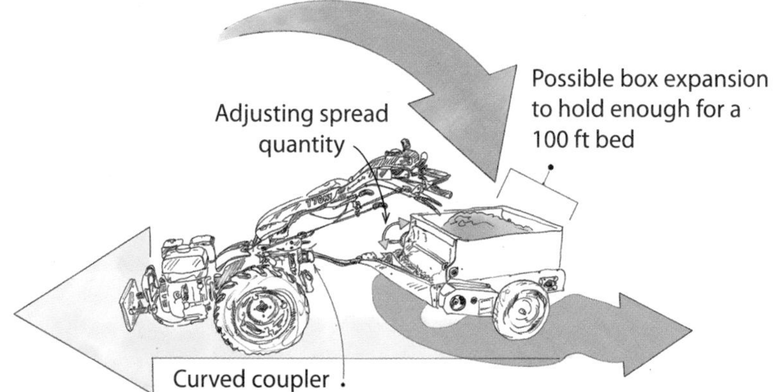

The spreader improves the evenness of routine compost applications in market gardens, orchards, and landscapes. Macro- and micro-nutrients can be applied as side-dressing by hand or with tools, but they can also be mixed into the compost for thorough and even applications with a compost spreader.

Pro Tip: *Mix your dry nutrient amendments into your compost and then apply liquid fertility with a backpack sprayer—or in-line with your irrigation.*

ADJUSTMENTS AND ACCESSORIES

This is a ground-driven pull-type implement used in rear-mount mode. Offset handlebars and walk adjacent to the spreader. A lever adjusts how much compost is allowed to spread out. An extension kit fits on the spreader box to increase load capacity for a 1" spread on a 100 ft bed. If the compost isn't coming out easily, stop and adjust the setting, *or the compost layer may be too thick.*

- **Accessories:** Small or large curved couplers are needed to fit different tractor models.

7. WINTER: SNOW THROWER (SINGLE- OR TWO-STAGE)

The snow thrower is a great way to add winter utility to your two-wheel tractor. The dual model has two stages with both an auger to gather snow off ground and center it toward the chute and a high-speed impeller that throws it out.

USE AND APPLICATION

The single- and dual-stage snow throwers clear snow where larger equipment won't fit, with widths up to 33". They add value to your two-wheel tractor by allowing you to get some winter use out of it. Snow throwers are great for short laneways, building edges, and paths. They can send snow 10–40 feet, depending on snow quality/quantity, terrain, and operational settings. But, as with all implements, you need to consider whether they are scale-suitable to your enterprise. ***Safety***: Be careful not to throw ice/rocks/debris, which can break windows and harm people.

ADJUSTMENTS

This is a PTO-powered push-type implement used in front-mount mode set in reverse to go forward. Adjust chute throwing direction with the control rod. Bottom skids set the height of snow intake, preventing biting into and spitting gravel/dirt/ice. Shear bolts protect the auger and impeller from damage when they break if the system clogs. Change if needed.

Maintenance tip: The snow blower has several grease nipples. Use a grease gun to apply grease to nipples for best maintenance of moving parts.

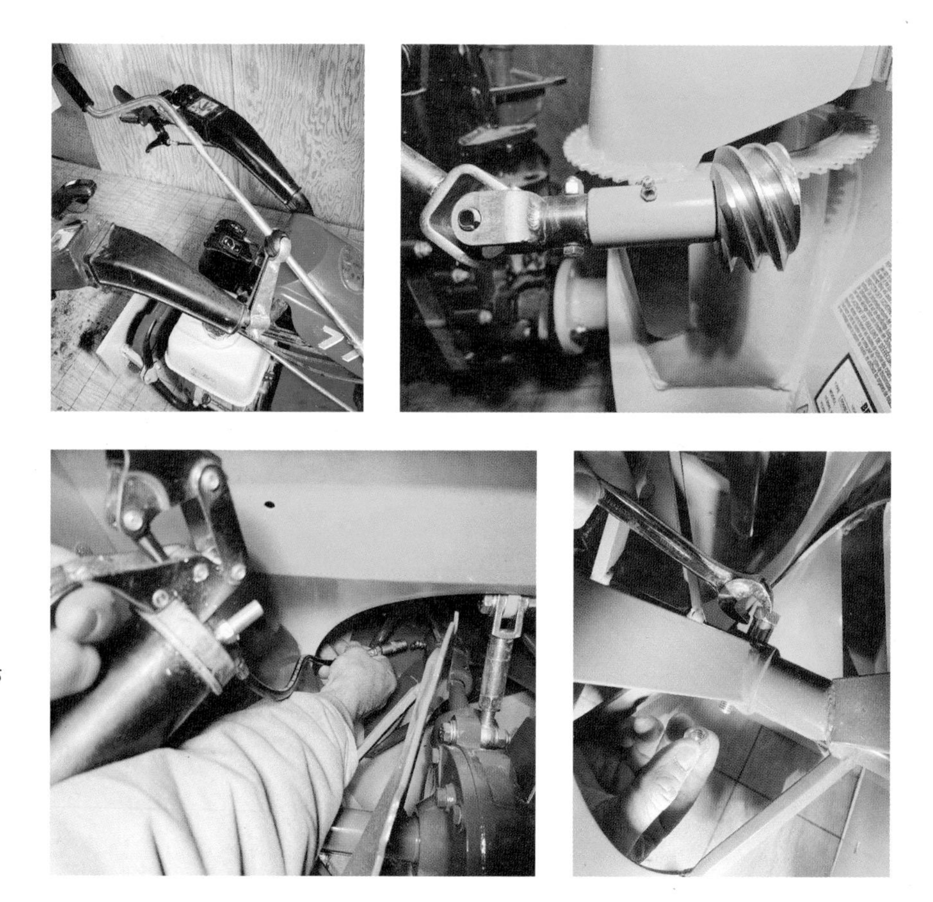

Chute adjustment rod is mounted on the handlebar.

Not all implements require this extra care, but for those that do, keeping your implement well-greased and knowing how to change a shear bolt is important.

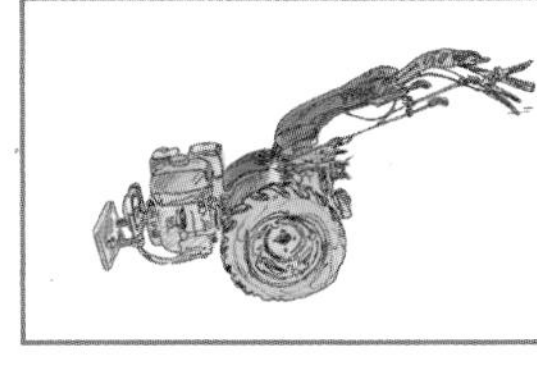

CHAPTER 4
Equipment Decision-Making

Remember, equipment decision-making is where growers can make the biggest mistakes or have the greatest successes. Having the right equipment can revolutionize your homestead or farm, but the wrong equipment choices can begin to dictate how you grow instead of facilitating your chosen production.

SCALE-SUITABLE EQUIPMENT

Is your bed preparation equipment the right width for your beds? Do you have the right type of tractor for your quantity of beds or rows? Equipment selection must be **scale-suitable** for best enterprise management. But of even more import is their suitability to your present and future scale. Finding the suitable scale is a balancing act; changes in crops, labor, or acreage affect equipment type, sizing, and quantity.

HOLISTIC PRINCIPLES OF SCALE

The **Holistic Principles of Scale** shows interrelated principles for land management decision-making. **Scale principles** include labor, acreage, extensive/intensive management, and more. All should be understood relative to one another. Use these principles when planning by asking yourself: *How does my consideration related to this principle affect the others?* For instance, equipment type is an important principle—with great variation. For some growers, hand tools are sufficient, while other farmers require two-wheel or four-wheel tractors. *How does this affect other principles of scale?* Using scale principles for decision-making can help determine which equipment is enterprise- and scale-suitable and focuses wise investment over time as you scale-up using multi-functional equipment strategies to get more bang for your buck!

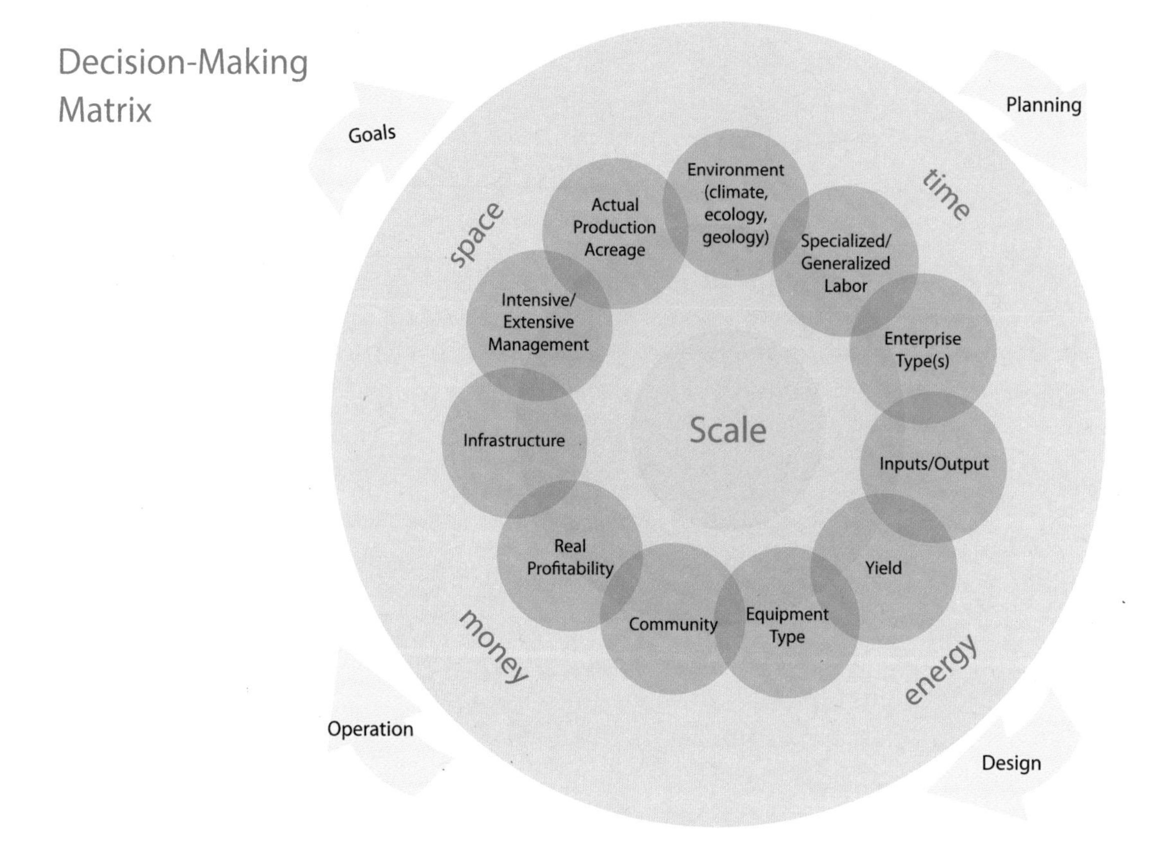

Decision-Making Matrix

ACTUAL PRODUCTION ACRES

One scale principle is **Actual Production Acres (APA)** of land that is *directly* managed with equipment (so, excluding buildings, wetlands, and other unmanaged land). This is between ¼ to 13 acres for two-wheel tractor operations, and usually 2–3 acres for market growing. If a Market Grower is considering working more than 2–3 acres, then they may need a two-wheel row crop tractor or even a 4-wheel tractor. Similarly, other scale principles, like labor, may need adjusting to account for a shift in APA.

SMALL-SCALE EQUIPMENT SPECTRUM

Small-scale equipment is a spectrum that starts with hand tools at one end, then wheeled tools, and, finally, two-wheel tractors. Four-wheel tractors are really in their own category, though they do overlap for some small-scale

DESIGN BOX

FOUR WAYS TO INVEST

Equipment type is one of many aspects of scale in balance as growers shift the relationships between **space, time, energy, and money.** These *four ways we invest* in projects change the balance of scale principles. For instance, increasing field management from intensive to more extensive is a *space investment* and will have savings elsewhere, like *time* from mechanized cultivation efficiency. This may require a new row crop tractor or a more powerful M2w-tractor, needing *energy (fuel) investment.* If we *invest money* in more types and larger equipment, we need to also consider if more infrastructure (*space investment*) for storage and maintenance is required. All of these are interconnected. Consider what your weak link is and invest accordingly, for some time is a limiting resource, and for other it is space.

land management. Professional and *larger* small-scale enterprises are often highly mechanized, with many different tools and implements—and often have more than one 2w-tractor. Knowing this can help you find which equipment-scale fits your needs.

Note: Growers using primarily hand tools and two-wheel tractors may benefit from a compact four-wheel tractor, especially for the use of the loader with bucket, pallet, and hay forks. But this decision should be taken with care, as it begins to mix equipment-scales.

Pro Tip: *Mixed Equipment-Scale Farming (MES) can be messy since differently scaled pieces of equipment are used together; this requires more know-how, investment, and consideration. However, MES is extremely common for farms over 10 acres and can be very effective—when jobs are assigned scale-suitable equipment.* ***Example:*** *Four-wheel tractors manage larger fields, and two-wheel tractors tend to be used for specialized jobs like greenhouse work, seeding, and cultivation.*

FIGURE 9: THREE EQUIPMENT-SCALES

For small-scale growers, there are *three main equipment-scales;* each has a crucial tipping point within scale-up. It's the weak links in production that will trigger equipment-scale shifts for certain tasks.

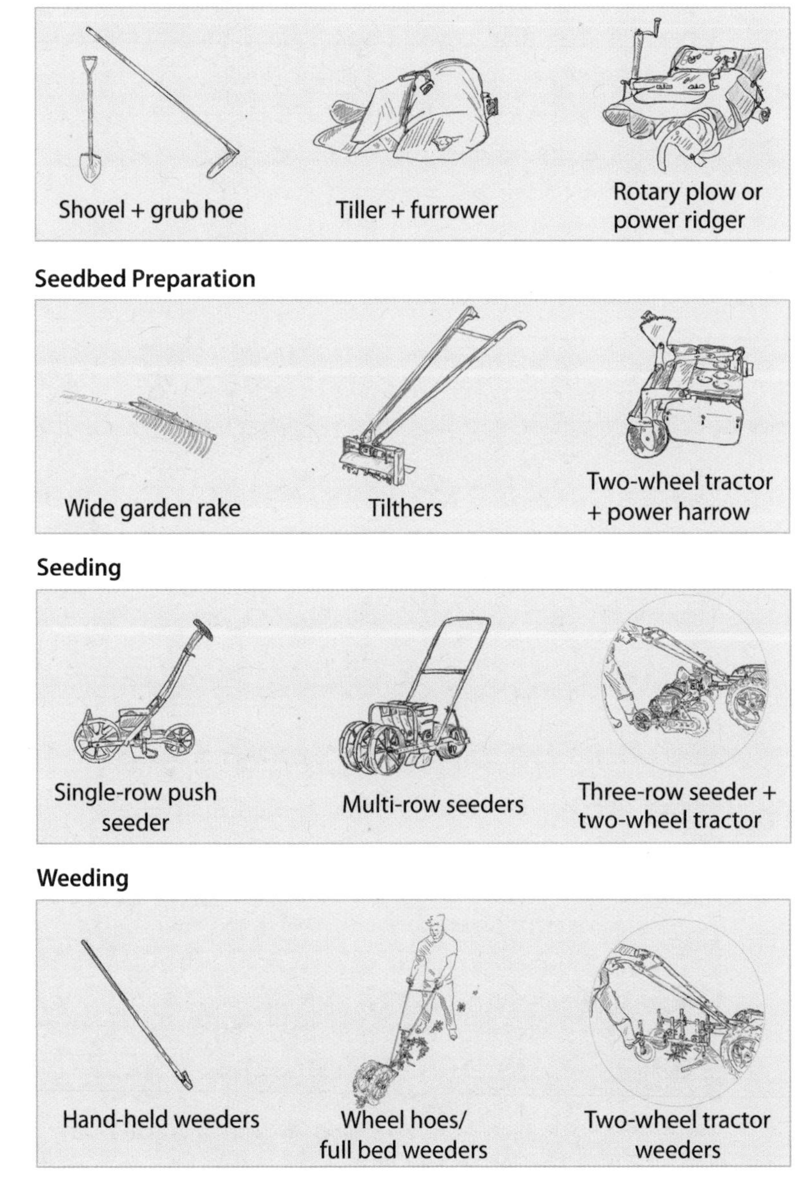

"We are ½ acre in raised beds and have 1.5 full-time equivalent workers on the farm. We seem to be around that tipping point where the investment in two-wheel tractors is paying off by doing enough work preparing beds."

— Josh Volk

In all projects, similar goals may be achieved employing different tools and equipment. Often, these complement one another, and sometimes they replace one another as you scale-up.

FARM FEATURE

CULLY NEIGHBORHOOD FARM AND SCALE-SUITABLE EQUIPMENT

Josh manages Cully Neighborhood Farm, a ½-acre urban operation in Portland, Oregon. The farm grows a diversity of crops for CSA customers using hand tools and a two-wheel tractor. The flail mower is used for garden edges, cover crops, and crop debris mowing. When the cover crop is mowed, the rotary plow is passed in the path, throwing soil over the mowed material which is left to decompose for 2–4 weeks. After this, the power harrow incorporates the material into the newly finished bed top as green manure. Josh uses a number of approaches to manage cover crops and bed turnover, experimenting with new methods to save energy and fuel.

For Josh, one question is whether their actual production acres justify two-wheel tractors. He has thought about going back to just hand tools—and

Photos Credit: Josh Volk

still may, one day—but ultimately thinks the "labor the tractor saves would never be completely paid for" at a tool-scale. If the farm used just hand tools, it would need to *trade off* its green manure strategies and perhaps opt to harvest the material standing from the beds and compost it elsewhere—requiring more *time and labor*. Or, the farm could rely on more imported material inputs, composts, and mulches to help manage bed preparation differently. The trade-offs come in increased costs for imported supplies, changes in equipment-scale (tools vs tractors), and the crop plan (bed turnover is quicker without the time needed for cover crop decomposition in situ). It is important to understand how multiple strategies can achieve similar results: healthy, nutrient-dense crops. But to get these results, a grower might use different equipment-scales. This exemplifies the idea of scale principles and how a shift in one aspect of your farm will affect others. It's an ever-evolving balancing act.

INTENSIVE/EXTENSIVE LAND MANAGEMENT

An important scale principle is ***intensive/extensive land management;*** its different aspects impact equipment selection. *Intensive* is partially defined by tighter row spacing and can lead to higher yields per square foot. *Extensive* refers to management that provides more space to crops, with lower yields per acre; it usually requires more equipment, but typically results in time and labor savings.

Intensive Example: A *Market Grower* often manages a 1-, 2- or even 3-acre garden *intensively*, with a two-wheel tractor making many passes per season on each bed, and with multiple crop/cover crop successions producing fresh market vegetables. Labor and input are high, but so are yields.

Extensive Example: A *Grassland Farmer* manages a 9-acre hay field *extensively* by cutting once or twice a year to provide hay for their dairy sheep to make artisanal cheeses. The value of 9 acres of hay is drastically lower than the production of a 1-acre market garden, but much less labor and fewer inputs are required.

Can you be extensive on small acreage? Yes, you can grow an acre of garlic in rows 6" apart (intensive), or the same acre of garlic can have rows 15" apart (extensive).

Does increased income justify more equipment? Ironically, intensive management's higher profit doesn't necessarily require more equipment, while extensive management usually does. The cost of intensive management is *usually* labor, imported nutrients, and rapid land-turnover techniques. Conversely, the cost of extensive management is space and equipment, but there is a savings in labor, and many inputs can be made in situ, like green manures and mulches, reducing supply imports (See: "Budgeting for Equipment Guilds," p. 123).

Haying is extensive *land management.*

DESIGN BOX
INTENSIVE VS EXTENSIVE

Intensive Features (per land unit):

- Closer row spacing
- More successions
- More operation passes
- Higher inputs
- Higher yields
- Higher labor needs
- Less and smaller equipment
- Often, higher-value crops, such as vegetables

Extensive Features (per land unit):

- Wider row spacing
- No or few successions
- Fewer operation passes
- Lower inputs
- Lower yields
- Lower labor needs
- More and larger equipment
- Often, lower-value crops, such as hay or grain

FIGURE 10: SCALE-SUITABILITY

Let's look closely at *scale principles:* **acreage, equipment, labor,** and **enterprise type.** The Urban Garden on .25 to 1.5 acre could do well with the BCS 722 with 1 to 3 full-time workers. They might opt for a professional model BCS 739 if they are Urban Pro Grower.

Full-Time Workers	1 to 3 (more for greenhouses and high-production models)	.5 to 6 (less for DIY gardeners, much more for professionals)	3 to 6 (less for routine work, more for planting and harvest days)	2 to 6 (more for orchards, less for hay)	1 to 6 (much less for DIY type enterprises, more for professional)	*The balance of equipment, labor, and enterprise types is part of the discussion of scale. Note: These numbers are meant to be relative quantities.

Actual Production Acres		Primary Enterprise Example
	Paths, Trails, Alleys, Lanes, Drives, & Roads	
~8 to 13 ac	Woodlot, Sugar Bush, and Whole Property	Back-to-the-Lander
~5 to 8 ac	Orchard or Hay Field	Grassland Farmer or Orchard Farm
~3 to 6 ac	Row Crop or Nursery Fields	Row Crop Farmer
~1 to 3 ac	Market Gardens or Homesteads	Market Grower or Homesteader
~.25 to 1.5 ac	Urban Gardens	Urban Pro Grower or Backyard Gardener

Number of and Examples of Tractors	1x BCS 722 (for Backyard Gardeners) OR BCS 739 (for Urban Pro Growers)	1x BCS 739 Add second tractor at about 2 acres	1x BCS 853 & PowerOx (m2w-tractor & row crop tractor)	1x BCS 770 1x BCS 739 (helpful to prevent haying time bottlenecks)	1x BCS 770 1x BCS 739 (for woodlots) or PowerOx (for storage crops on Back-to-the-Lander properties)	*Very Production Focused Enterprise may even add a 3rd tractor at larger acreage scales.

The limit of two-wheel tractor suitability is not set in stone. There is no magical "acreage-ceiling." That being said, one can get a general sense of the equipment, labor, and actual acreage needs for different enterprises.

FARM FEATURE

TOURNE-SOL CO-OPERATIVE FARM IS MIXING UP EQUIPMENT-SCALE

It is possible to grow using different equipment-scales. Tourne-Sol Co-operative Farm outside of Montreal, Quebec, operates with a "mix of extensive techniques and intensive production," says Reid Allaway who is in charge of all things tractor and equipment on the farm. They use four-wheel tractors but are increasingly applying two-wheel power. Their BCS tractor does greenhouse bed preparation for high-value intensive crops, but a two-wheeled Planet Jr does 75% of the crop seeding. This saves time/labor because four-wheel seeding required a second person to walk behind to ensure the hopper had enough seed and the seed-furrower wasn't jammed.

Now, they have taken their two-wheel tractors electric, which has lowered operating costs, eliminated emissions, and reduced noise. This electric two-wheeler is great for greenhouse use as there is no reduction in air quality from emissions. Reid says the Planet Jr is the easiest type to convert, costing about $600 in parts and 25 hours (for a skilled person who likes to scrounge for parts).

"Battery-electric power is definitely the way forward for walk-behinds, and there's no major technical hurdle that can't be easily cleared with readily available tech."

— Reid Allaway

PHOTO CREDIT: GHISLAIN JUTRAS – L'ODYSSÉE BIO DE GIGI

PHOTO CREDIT: REID ALLAWAY

Reid Allaway talks cultivation options for his fixed-up electric-converted Planet Jr BP-1.

SCALE PHASES

Projects grow over time. When you plan your start-up with clear intentions for future enterprises and production acres, you can nurture *strategic* growth. Pitfalls of *unhampered growth* include growing beyond your goals and poor equipment investment choices. When you purchase based on pressing problems, you usually end up with equipment that is not complementary or equipment that has rapid obsolescence. Hasty equipment choices may *dictate how you grow* instead of being *chosen for your goals*.

THE EVOLUTION OF SCALE

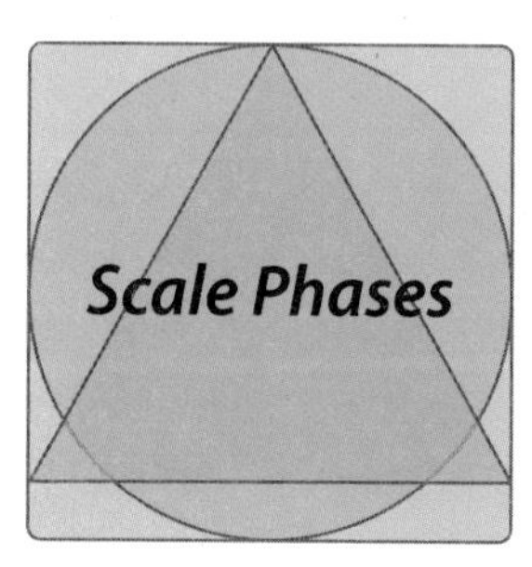

All land-based enterprises go through **scale phases:** from start-up, to scale-up, and finally pro-up. Understanding that there *will be* scale phases is powerful knowledge to have for budgeting and planning. Examples of scale phases include the process of expanding in acreage, equipment, and profit to become well-established, or a DIY-grower starting up, scaling up, and eventually *reaching and transitioning to* a professional enterprise at the pro-up phase.

- At **start-up**, growers focus on goal-setting and *essential* production solutions.
- With **scale-up**, growers focus on planning additional projects and productivity.
- At **pro-up**, there is a refining of operation cycles and solving of remaining weak links.
- At your **static scale**, focus is on continued steady-state operation!

PRO-UP TO A STEADY-STATE

The three phases should always be seen in relation to your goals and include an *intended* **static scale**—where growth is limited. Your enterprise will grow, but it should stop after the pro-up phase, at which point aspects of your operations, such as the amount of acreage, acquisition of equipment or adding new labor, should stop growing. But since there are always minor shifts from year to year—maybe you grow fewer melons if demand is lower, or maybe you add one extra fieldhand in a year with a bumper crop of cherry tomatoes—but ultimately your production is similar in quantity of land and the quality of production, and the equipment and techniques that you use from year to year. We refer to this management as *steady-state* (meaning small fluctuations, but overall keeping at the same scale). This steady-state management at your static scale is reached after pro-up, when the *intended* balance of acreage, equipment, profit, and enterprise types is achieved! All these aspects of production grow and expand during scale phases but no longer grow after the pro-up phase and are instead operated at a static scale with some minor growth or reduction (steady-state) from year to year.

SETTING A STATIC-SCALE GOAL

An enterprise's **static scale** is the point of maximum acreage, equipment, and most other growth investment. This is the point at which all scale principles are in relative balance, and management is in a steady-state equilibrium. Once achieved, adjustments only need to be made *for unique circumstances or for specific changes in the farm's overarching goals.*

Land stewards must set **static-scale goals**: acreage, profit, labor, etc. This constrains our designs and planning, revealing correct choices for purchases and equipment-use strategies. Intended static-scale goal-setting helps avoid common pitfalls of ad hoc expansion and equipment acquisition—before a property design and business plan are fully formed. All too often, I've seen an unhampered acquisition of "heavy metal" weigh a grower down and make them less flexible to changes in their business plan and enterprise goals. You only get where you want to go by *saying* where you want to go, and this means setting a scale goal. Consider the example static-scale goal below, and remember to always state goals as an affirmative and to clarify the boundaries of your scale phases!

Pro Tip: *The more precisely you define your static-scale goal, the better choices you can make for equipment as you start-up and scale-up.*

DESIGN BOX
DEVELOPING A STATIC-SCALE GOAL

Michael (23) and Anna (24) want to start a small-scale farm. Here is their goal, so far:

"We will manage a 2-acre market garden, working full-time for 8 months and part-time for 4 months, with three seasonal workers. We will buy our land with a loan at 4% interest over 30 years, repair the 1,000 sq ft barn, and put up a 20 ft × 30 ft greenhouse and a 16 ft × 100 ft caterpillar tunnel. We will grow on 200 4 ft × 100 ft raised beds arranged in plots. We plan to sell organic vegetables at farmers markets and on-farm. Our planned gross sales are $225,000 to $300,000 at our intended static scale. We will use intensive techniques, like green manures and zipperbeds using hand tools and two-wheel tractors. We anticipate start-up investments for equipment and infrastructure in years 1 to 3 to equal to 50% of intended gross sales; the start-up investment will be met through grants and savings, and 5 to 10% of annual gross will be allocated for equipment to scale-up over years 3 to 6. In years 6 to 9, we will allow our operation to streamline and refine our management rather than expand, with less equipment purchase and more calibration. Our static scale will be reached with a balance of all principles under management: acreage, equipment, profit, labor, etc."

Now, this is a *very clear picture* to frame decision-making and bounce equipment types and operation ideas against!

When your intended static scale is identified early on, higher-level equipment might be acquired sooner and be able to perform basic functions at start-up as well as those functions anticipated for later, at pro-up. For instance, when unclear if a backyard garden will become a market garden, then build raised beds with the affordable **Basic Garden Guild**. (A discussion of equipment guilds will follow shortly.) Conversely, if your scale goal is clearly a larger market garden, then go for the **Land Prep Guild** from the get-go.

BUDGETING WITH SCALE PHASES

Scale phases allow better planning and budgeting, including finding seasonal deals on equipment, sourcing them used, and taking time to consider accessories and options. Setting a static scale doesn't mean your plans can't change, but it certainly makes a smooth road map to follow, making decision-making pit stops easy-breezy.

EQUIPMENT CONSIDERATIONS FOR SCALE PHASES

Let's look at equipment considerations for each scale phase for garden-based enterprises.

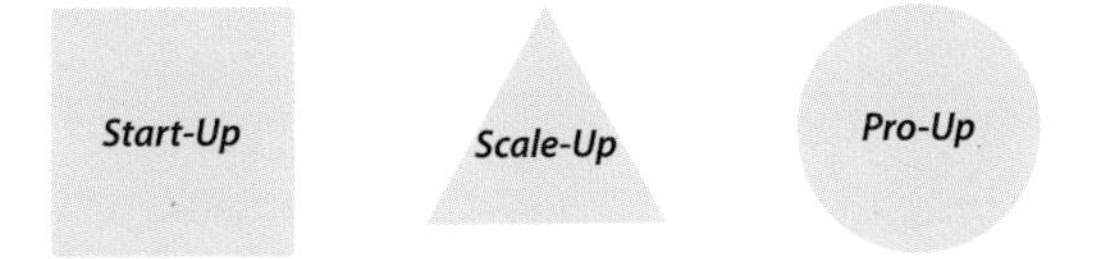

1. **Start-up** equipment considerations for garden-based enterprises are marked by the initial need to transition lawn or old fields into production acres. Scale-suitable equipment is focused on garden maintenance and enhancing methods used for essential operational cycles, such as bed forming, seedbed preparation, and crop debris management. Tillers and rotary plows are typically used. Hand tools are essential for cultivation, harvest, and post-harvest packing. Some growers start with *just* hand tools!
2. **Scale-up** equipment considerations are marked by the need to increase production by using additional mechanization and tackling more mechanized fertility and harvest efficiency. Compost spreaders and a cart for harvest may be needed. A power ridger will make bed reforming a breeze. Also, additional enterprise(s) and/or specialization within the original enterprise often occur, and new equipment is selected and multi-functional uses are explored for current implements.
3. **Pro-up** equipment considerations should be marked by the aim of reaching a static production scale at maximum production acres with a complement of multi-functional equipment to meet all enterprise needs. Highly refined and efficient operational cycles are the goal. This phase fine-tunes equipment and focuses on weak links. Cultivation and seeding equipment may be needed. Investment in post-harvest processes and marketing is typical. New techniques and equipment can be added, but only as thoughtful innovations, not willy-nilly changes in the steady-state system.
4. ***Note:*** Sometimes, farms transition to new land or new production goals, and these phases (to some extent) begin anew. But we can learn how to apply our equipment and skills to new contexts!

FARM FEATURE

EQUIPMENT AT AZIENDA MONTE REALE

Serena's farm in the Ligurian Hills of Italy has small plots on either side of a winding road that leads to the mountain top and farm center. She returned to reinvigorate the family farm eight years ago. She now grows vegetables, cultivates "thousand-flower" honey, and raises goats, geese, and chickens. Her honey is full of wildflower and chestnut flavors. She also makes specialty products, such as lavender bundles for holiday markets.

She started as a Backyard Gardener with a scythe, a wheelbarrow, and a rake. She then brought in a powered wheelbarrow and a rototiller. As she **scaled-up,** she replaced the rototiller with a two-wheel tractor (a BCS 738) and saw the multi-functional power of this equipment. Her first implement purchases included the rotary plow, rear-tine tiller, and flail mower. She then added a sickle bar mower and root digger to her repertoire. She also has a wood chipper and many hand tools. This project is in a **pro-up phase,** fine-tuning equipment for her **static scale.** For Serena, "each tool is of fundamental importance to save a lot of time and, above all, effort."

Left side photos credit: Serena Balbi

EQUIPMENT GUILDS AND OPERATION CYCLES

Equipment decision-making can be helped by the design of **equipment guilds. A guild** is a *companionship of three or more mutually beneficial entities.* There are guild designs for plants, equipment, and enterprises. Equipment guild design conceptualizes equipment as working together *as a unit* to complete an operation cycle. An **operation cycle** is a series of tasks that always go together and must be finished for successful completion of a specific stage of growing. For instance, the **Land Prep Guild** (rotary plow, rear-tine tiller, and flail mower) can open new ground, and build, finish, and reform raised beds or Permabeds.

Pro Tip: *Equipment guilds are assemblies of multiple implements that help complete an entire operation cycle; they are used to strategize equipment acquisition, multi-year budgeting, and multi-functional use while scaling-up for specific enterprises.*

WHY EQUIPMENT GUILDS?

Equipment guilds put the grower in the mindset of "which implements work well together" to complete *whole operation cycles* rather than focusing solely on specific tasks. An implement can be part of one equipment guild and also be used in another for a different task. This makes the implements *multi-functional.* In this way, growers can strategize to bring in implements that work as a guild in their **start-up** phase to complete entire operation cycles, and then pair some of these implements with new equipment purchases in **scale-up** and **pro-up** phases to form other equipment guild combinations.

DESIGN BOX

WAYS TO USE EQUIPMENT

Equipment can be organized singly, by their **specialized tasks,** or sorted into **equipment guilds** that complete **whole operation cycles**—a series of tasks as a unit. Then, there is equipment that can be used for different tasks as part of different operation cycles (multi-functional).

1. **Specialized task:** Plowing a field for the first time or seeding are examples of specialized tasks.
2. **Whole operation cycle:** PDR tiller breaks up plow ridges, rotary plow forms raised beds, and power harrow finishes bed tops. This completes the operation cycle of *forming new beds.*
3. **Multi-functional use:** This is when equipment is employed for different projects, performing many tasks within many operations cycles.

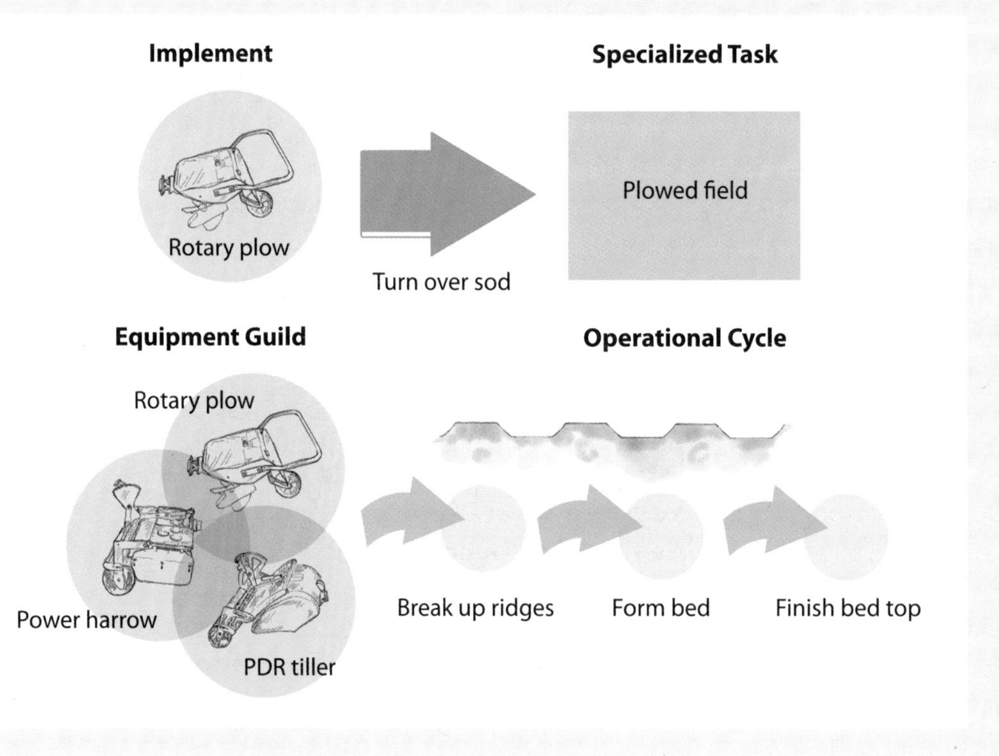

EQUIPMENT GUILD DESIGN

The equipment guilds presented here have been designed strategically, and their names hint at their utility. The Land Prep Guild is named for implements that prepare land for gardens, but the guild is also helpful for general land preparation for orchards, landscaping, reforestation, etc. Conversely, the Homestead Guild contains equipment particularly useful to DIY property owners.

Equipment guilds are designed for:

1. **Low-risk:** Equipment guild recommendations for specific enterprises are generally useful for most people managing that type of enterprise. For example, the Basic Garden Guild is a good choice for most small Backyard Gardeners.
2. **Completing operation cycles:** A guild's implements work together to complete specific work, such as land preparation or cultivation. When scaling-up, equipment purchases are staged to spread out costs while still prioritizing completed tasks at each scale phase.
3. **Enterprise-specific:** Organized by the type of enterprise, equipment guilds are specifically associated with a management style and enterprise type, such as the Homestead Guild, which has a unique blend of practical, four-season equipment for Homesteaders.

EQUIPMENT PURCHASE PLAN FOR 3-ACRE GROWER

As discussed, when a future static scale (say 3 acres) is known, then equipment suited to multiple phases can be selected at start-up and incorporated easily in later phases. For instance, picking up a tiller with a 30" width initially—even if 26" would be fine for start-up—is a smart move—if you know your future intended scale will benefit from the 30" model.

Our 3-acre grower will purchase the **Land Prep Guild** at **start-up** to effectively open new land with a rotary plow and prepare fine seedbeds with a rear-tine tiller. Conversely, selecting the **Basic Garden Guild** would not make sense, knowing the intended growth of actual production acres. Then, the grower can **scale-up** with a **Soil-Building Guild** (power harrow, power ridger, and compost spreader) to mechanize soil conservation methods. This would be followed with a **Row Seeding/Weeding Guild** to **pro-up** for efficient cultivation of increased acreage later for some cash crops (such as winter storage carrots) or invested in other equipment fine-tuned to their final goals.

FIGURE 11: SCALE PHASES

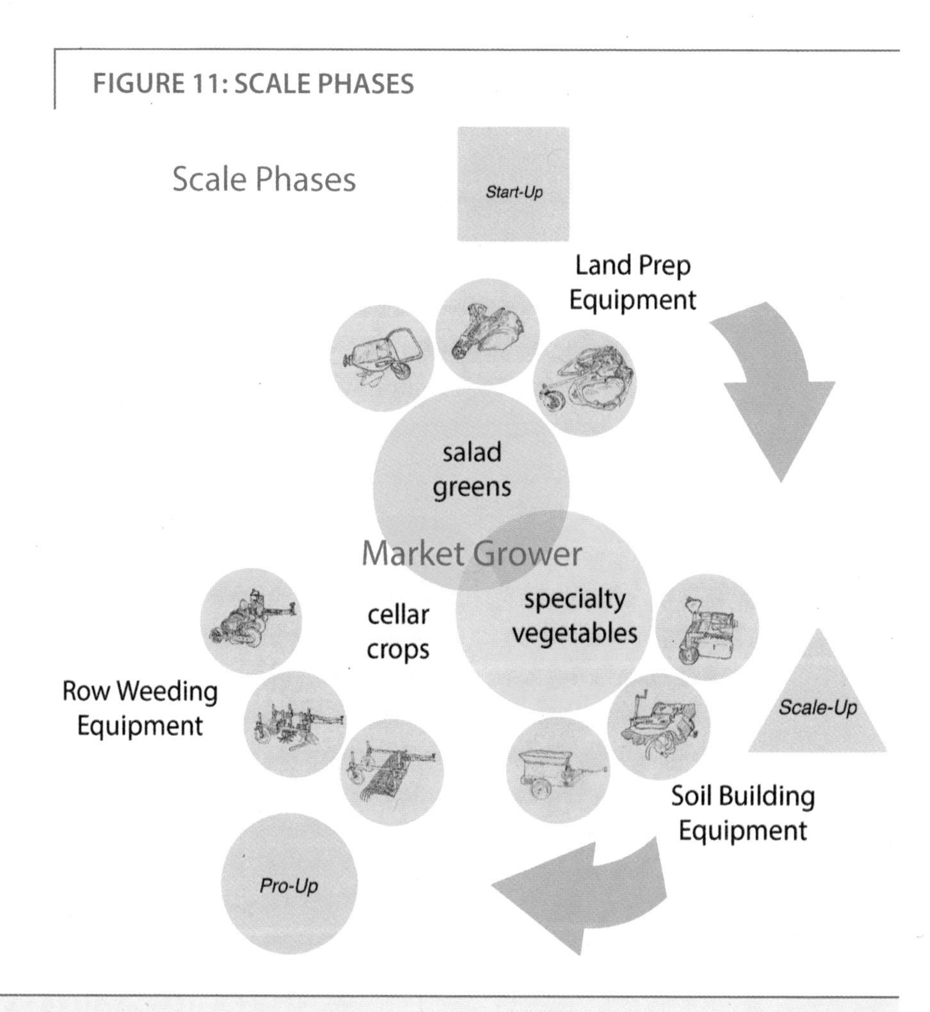

DESIGN BOX

MECHANIZE FOR GREATEST NEED

Consider the grower's intention for 3 acres in a market garden. The goal is to mechanize for efficiency of operations, in order of greatest need. 1) Opening and preparing land is labor-intensive. Think ahead. More land than is needed should be transitioned to garden—years in advance, whenever possible. 2) Mechanize compost and cover crop activities. 3) Finally, with full acreage in vegetables, mechanize weeding. ***Note:*** Last of all, growers will need to focus on investment in the mechanization of harvest and post-harvest equipment (which are not two-wheel-tractor related—but will be featured in an upcoming book!).

EQUIPMENT GUILD EXAMPLES

BASIC GARDEN GUILD

This set of equipment provides five key seasonal functions at a low initial cost: preparation of new beds, reforming of garden beds, fine seedbed preparation for seeding, mowing to clear land for the new garden plot, and cost-effective cover crop management. You can add on a cart to replace your wheelbarrow to help with fertility additions and harvest loads. Employing this guild is a great way for Backyard Gardeners or Market Growers to start up.

- **Rear-Tine Tiller (26", or similar)** is used to open previously worked ground (often the case when planting into suburban lawns), and it prepares seedbeds for planting.
- ***Note:*** Rear-tine tiller *cannot* be used to open hard-packed soil or fill.
- **Hiller/Furrower attachment** for the tiller is a low-cost solution for forming paths and effectively designating and building up low raised beds. It can also furrow a trench for planting potatoes, asparagus, etc.
- A **Brush, Combo, or Flail Mower** is used to mow crop debris and cover crops and for basic lawn management. For a Backyard Gardener, the combo mower is a good choice because it has the right price point and serves multiple functions.
- **Add-on:** A **Buddy Cart** is handy for bringing supplies to the field in spring (compost), carrying harvests in summer, and removing debris in fall.

The Basic Garden Guild is a modification of the original low-cost, low-risk essential-three implements, with the added buddy cart replacing the ubiquitous wheelbarrow for some tasks

THE ESSENTIAL-THREE EQUIPMENT

The well-known *essential-three* start-up equipment—rotary plow, flail mower, and power harrow—are actually a recent phenomenon, with availability in North America starting around 2008. The *original* essential-three (available since the 1970s) was a rear-tine tiller, a hiller/furrower, and a bush hog or lawn mower. The opening of new land was often outsourced to a neighbor with a 4-wheel tractor, and then the Market Grower took over the earthworks for bed preparation.

These *essential-three* were an affordable start-up equipment option to complete essential garden tasks: bed forming, seedbed prep, and crop debris management. It was with these that I learned to market garden in New Mexico, and it was with the affordable hiller/furrower that I built my original Permabeds on a 1-acre market garden in Ontario.

This equipment guild is still a budget-wise, low-risk farm start-up option. Consider the affordability: 26" rear-tine tiller ($940), hiller/furrower ($105), and 26" brush mower ($1,490) with a total price of *only* $2,535. Even if you added the buddy cart ($785) to replace the popular wheelbarrow, and modernized the mowing by upgrading the brush mower to the very versatile 30" flail mower ($2,725), your start-up implement cost would still only be $4,555. Even with the kick-ass BCS 739 ($4,545) (we had a 725 in New Mexico, which had no differential, making turns in small plots a bit tricky), you are talking under $10,000 for all the mechanized equipment necessary for serious garden or market garden start-up.

With this low start-up investment, growers can then add other essential equipment at the 1- and 2-acre tipping points to improve efficiency. Examples: adding a power harrow to improve soil conservation, or getting a rotary plow or power ridger to scale-up bed reforming. The other equipment is still useful too; I continue to use my furrower for planting asparagus, flower bulbs, and more.

FARM FEATURE

GEMINI FARM

Gemini Farm in New Mexico was an early pioneer in two-wheel tractor farming. They did all their essential work with a bare-bones Basic Garden Guild: tiller, furrower, brush mower, wheelbarrows, shovels, and rakes. They pushed the guild to the limit at 3 acres in organic vegetables. At that point, they scaled-up by adding additional BCS tractors. After that, they moved to a mix of mule and M2w-tractor power. They made a customized bed former that consisted of a rear-tine tiller with its outer tines and side guards removed, with the addition of two furrowers welded together to make a double-wide path trencher that could also dig deeper. ***Note:*** I am not recommending this—but it was an innovative and low-cost solution and is part of the story of how growers are *always* leading the changes in equipment for a new agriculture. The power ridgers now available do a great job of safely forming beds. They are a great investment for serious raised-bed growers.

LAND PREP EQUIPMENT GUILD

This is an essential guild for a Market Grower's (1–3 acre) start-up and indeed is considered to be one of the best start-up configurations for many enterprises, including Urban Pro Growers working as little as ¼ acre but needing fast garden bed turnover. This is the *new* essential-three implement guild: rotary plow, flail mower, and rear-tine tiller (although some growers opt for a power harrow over the PDR rear-tine tiller). If you are serious about market growing, choose this equipment guild over the Basic Garden Guild from the get-go. For growers with 1–3 acres, this is the go-to for start-up.

- A **Rotary Plow** is used to open land and form raised beds. What the rotary plow does differently from the hiller/furrower is actually breaking and *then* displacing soil, as opposed to simply displacing. This is why it is now the implement of choice for Market Growers who invest earlier in better equipment.
- The **PDR Rear-Tine Tiller** (30", or similar) is used for breaking up plow ridges, and then, with the added *Precision Depth Roller* (PDR), it helps in making finished bed tops by pressing the newly turned soil to make a firm seedbed for better soil-to-seed contact and improving germination.
- A **Flail Mower** is used for managing crop debris and cover crops.
- **Add-on:** A **Subsoiler** is used to break up hardpan and open the soil for water and root penetration. It is especially useful when opening new land initially, such as old clay-based pastureland.

The "new" essential-three equipment.

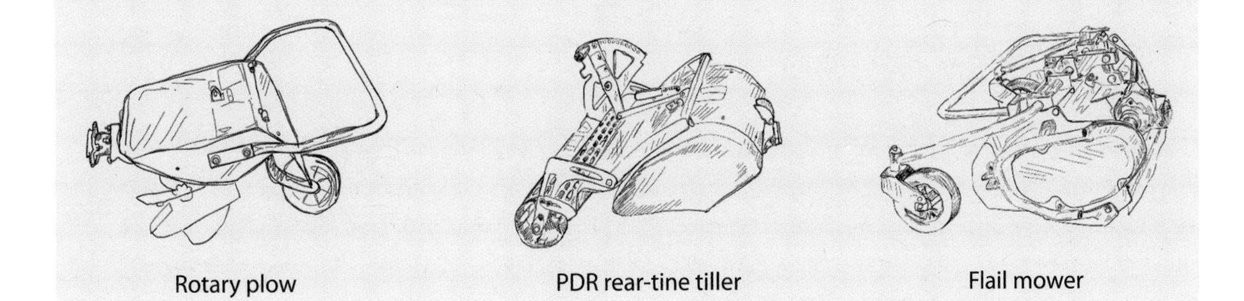

HOMESTEAD EQUIPMENT GUILD

This equipment guild adds some great implements to take you from backyard growing with the Basic Garden Guild to full-blown homesteading. This set of equipment allows you to diversify your potential undertakings, including mowing on larger acreage, providing irrigation, and doing winter snow blowing.

- A **Lawn Mower** (or other mower) is used to maintain lawns, walkways, garden alleys, and an under-orchard canopy. Choose a lawn mower if you have more flat ground to mow; choose a flail mower if you are mostly focused on cover crop management; and choose a combo mower for a bit of both.
- A **Single-Stage Snow Thrower** is used to maintain edges of buildings, small driveways, and walkways. If you have deep average snowfall or heavy drifting, consider the two-stage model.
- A **Water Transfer Pump** is used for small irrigation jobs, removing unwanted localized flood water in ditches and even filling/emptying small ponds or pools. For small Homesteaders, a pump can be attached to your tractor's PTO and used to transfer water from storage tanks collecting roof water for irrigation.

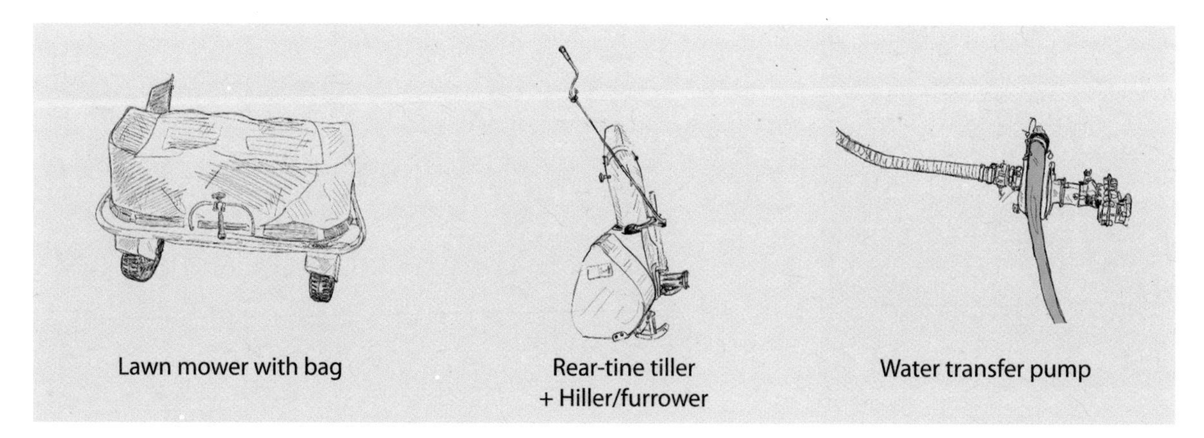

This equipment guild can help with many varied tasks—a necessity for the Homesteader.

SOIL-BUILDING EQUIPMENT GUILD

This equipment brings fertility management to an efficiency needed for scaling-up many land-based enterprises: gardens, homesteads, orchards, afforestation, etc.

- A **Power Harrow** is used to prepare and re-prepare fine seedbeds. This is a soil conservation tool because its vertical tine rotation doesn't bury topsoil and won't create a plow pan from the tine "slapping" action of a tiller. For growers making frequent passes when managing intensive market plots, this is a go-to for scale-up. A bumper weight is recommended for most tractor models.
- A **Compost Spreader** is used to apply *finished* compost to a bed top as *initial* and *seasonal* fertilization. A 1" spread is the standard seasonal application; more is better, though, when starting a garden from scratch. This implement replaces the wheelbarrow and shovel typically used at start-up.
- **The Power Ridger** is used to reform beds efficiently by distributing path soil onto the bed top of two adjacent beds. Using the Compost-a-Path Method (see Chapter 6 "Compost-a-Path Principles"), this equipment is key to creating in situ compost. Also helpful for any type of trenching, from planting potatoes to burying irrigation lines.

Power harrow Compost spreader Power ridger

Pro Tip: *Some growers (with under 2 acres) may opt for either hiring more labor and continuing the use of wheelbarrow fertility application or jumping right in and buying a compact 4-wheel tractor with a bucket for compost applications (especially useful for growers with more than 3 acres). The compost spreader is for those who want very consistent compost top dressing and/or for those who prefer low-cost mechanization rather than new labor hires or an entirely different equipment-scale.*

ROW SEEDING/WEEDING EQUIPMENT GUILD

This equipment mechanizes seeding and in-row/between-row cultivation for extensive row crop systems. This is a great pro-up equipment guild, especially for Market Growers who want to add some extensive crops to their more intensive market garden. However, full-blown Row Crop Farmers will have *different* equipment guilds for their scale-up phase since they will have different priorities when all the land is managed extensively.

- A **3-Row Seeder** is used to seed crops at a precise depth and equidistant row spacing. The Planet Jr or Jang 3-row seeder are two options. ***Note:*** Some growers opt for a multi-row hand seeder (like the Jang) in place of a tractor-mounted one.
- A **Tine (or Wire) Weeder** is used to pre-weed (also called *false sowing*) the bed top before seeding crops, and then blind weeding the crop rows when weeds are at the white thread stage.[1] (See the Design Box: "Understanding Tillage and Cultivation.")
- **Finger Weeders** and **Tender Plant Hoes** are used to effectively weed in-row and between-row weeds when crops get larger. Finger weeders are made of flexible rubber fingers that dip in-row to remove weeds, and tender plant hoes can cut close to crop rows. Neither tool will harm crops.
- **Add-on:** Some growers add a basket weeder or swap it for the tine weeder. It won't blind weed, but it will do an excellent job of pre-weeding and doing tight weed management of tender crops like carrots (where too much soil kicked up into the row affects young seedlings).

1 This is a name for very young weeds that look like white threads at an early stage of growth. They can sometimes be cultivated at a size that is almost imperceptible to the "uncultivated eye."

- ***Note:*** Gauge wheels are usually needed with cultivation implements to help set the depth and to control the movement of the implement.

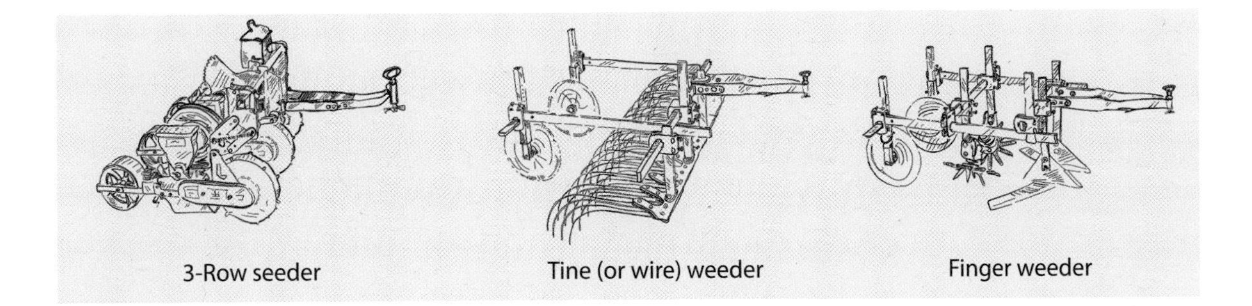

Pro Tip: *Multi-row seeders are for growers who will routinely seed three or more 100 ft beds of the same crop. For instance, I used to succession-seed 3 to 6 Permabeds every two weeks with carrots: early baby carrots at 5-rows, then summer and storage carrots at 3-rows. Carrots are a very lucrative cash crop, and 3-row seeding, pre-weeding (false sowing), and good cultivation were key on my farm.*

There are many different cultivator arrangements for row crop growing, but as Market Growers pro-up, a 3-row seeder, tine weeder, and finger weeder will cover most of their bases.

TRANSPLANT EQUIPMENT GUILD

This equipment helps Market Growers manage greater numbers of *transplanted crops,* helping to form and prepare mulched (transplant-ready) garden beds. This is an option for some growers using mulch for transplanting. In this way, beds are raised, and a synthetic roll-out-type mulch is used and holes punched through it for planting.

- A **Bed Shaper** is used to raise up a bed; it works very precisely with a mulch layer accessory (usually about 3" tall).
- A **Mulch Layer** is used to roll out poly or biodegradable mulch and can include an optional drip tape layer. ***Note:*** Consideration of mulch material is needed to make sure it works with your operation and ethic.
- A **Rolling Dibbler** is used to place precise holes in the mulch for planting. This tool is also useful for helping plants into bare soil, as it makes sure the planting is precise—especially useful for crops like garlic.

- **Add-on:** Pull-type dibblers for the two-wheel tractor can be made from scratch or modified from the waterwheel transplanter used for four-wheel tractors.

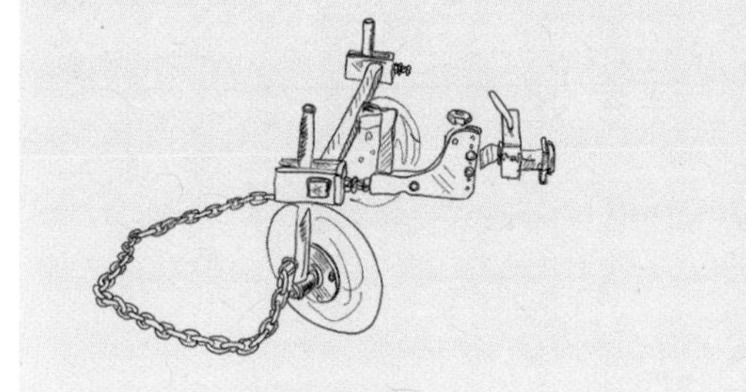
Bed shaper

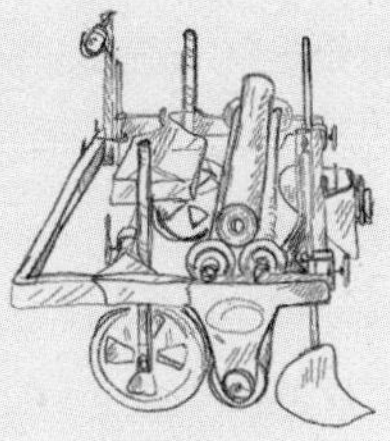
Mulch layer

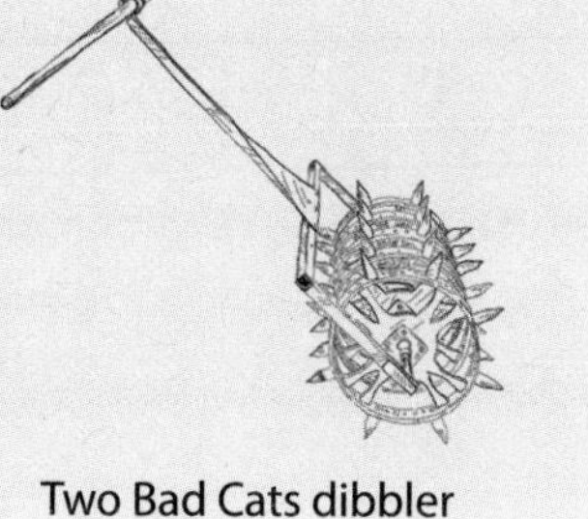
Two Bad Cats dibbler

TREE PLANTING GUILD

This equipment covers all basics for serious earthworking for the management of trees.

- A **Power Ridger** is used to form trenches for planting fruit trees.
- A **Rotary Plow** is used to fill trenches around trees for nurseries, reforestation, or orchard planting.
- A **Power Harrow (PDR tiller)** is used to soften edges of tree rows for cover cropping and intercropping and for preparing bed tops for multi-row beds.

Power ridger

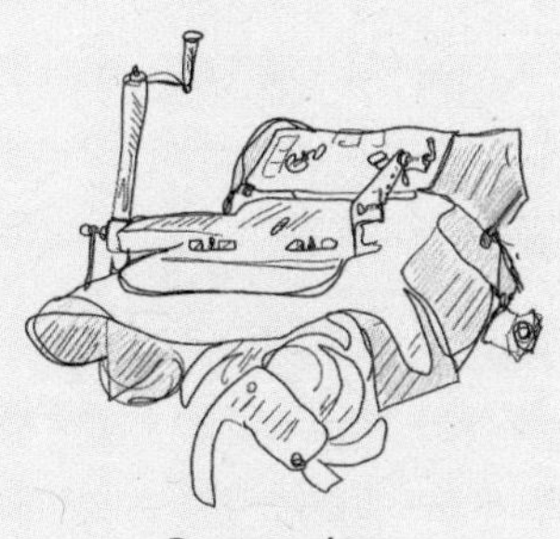
Rotary plow

Power harrow (PDR tiller)

FORESTER EQUIPMENT GUILD

This is a great enterprise-specific equipment guild for DIY enterprises looking to manage woodlots.

- A **Log Splitter** splits woodlot logs into firewood for homestead heating or sugar shack use.
- A **Chipper/Shredder** chips woody material into pieces that can be used to define paths or mulch fruit or maple trees.
- A **Utility Cart's** flat bottom makes it useful for hauling sap buckets (5-gallon pails with pressure-fitted lids), and its dumping feature makes it ideal for moving supplies like chip mulch and wood out of the woodlot.
- **Add-Ons:** A **Brush Mower** is useful for clearing thick-diameter weedy plants and small shrubs. A **Flail Mower** is great for maintaining paths within woodlands.

Log splitter (bottom left) The power cradle (bottom right) is an essential accessory for a pressure washer or water transfer pump (though NOT for an irrigation pump) and log splitter.

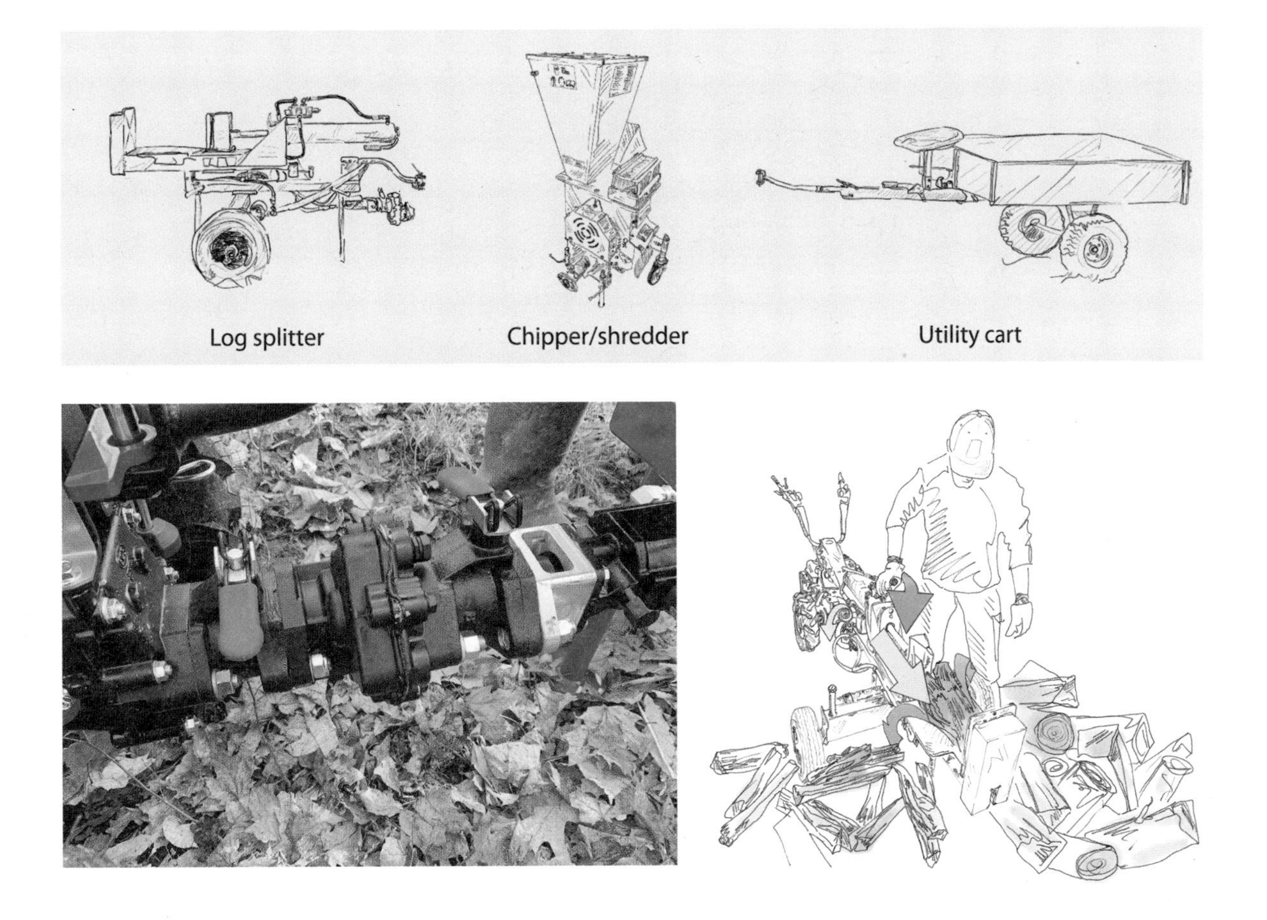

Pro Tip: *Chip-prune orchard wood into ramial chipped wood (RCW) for fertility management in gardens and orchards. RWC, composed of fresh wood chips from small-diameter saplings, is much higher in nutrients than older wood. RWC can increase soil organic matter, stabilize carbon, and improve soil ecosystem health.*

FOUR-SEASON EQUIPMENT GUILD

This is another specialized equipment guild meeting a slew of odd jobs for serious Homesteaders and Back-to-the-Landers.

These essential implements help property owners with many projects balance other equipment, infrastructure, and landscape needs.

- A **Pressure Washer** keeps vehicles, infrastructure, and equipment clean. **Examples:** root cellar clean-out in spring, spraying off the farm truck, and cleaning an old barn for a new purpose.
- A **2-Stage Snow Thrower** removes snow for farm and homestead lanes and around infrastructure.
- A **High-Pressure Irrigation Pump** provides water for irrigation. Back-to-the-Landers will have many needs for small-scale irrigation: gardens, orchards, crop fields, and more.

Pressure washer

Snow thrower

High-pressure irrigation pump

HAYING EQUIPMENT GUILD

- A **Sickle Bar Mower** cuts grass to be dried and raked as hay. A dual-action model is helpful to get big cuts in one pass and reduce operator fatigue.
- A **Hay Rake** is used to rake hay into a windrow once it has dried properly. This can also double as a tedder, to shift hay if it needs to dry out more and to dry hay out again if there is a surprise rain.
- A **Round Baler** is used to collect the hay windrows and roll them into tight, round bales and wrap them with synthetic bale wrap mesh.
- **Add-on:** A **Compost Spreader** is used to distribute compost into the hay fields to enrich them with fertility. ***Note:*** The compost spreader *is not* a manure spreader and *cannot* handle bulky material.

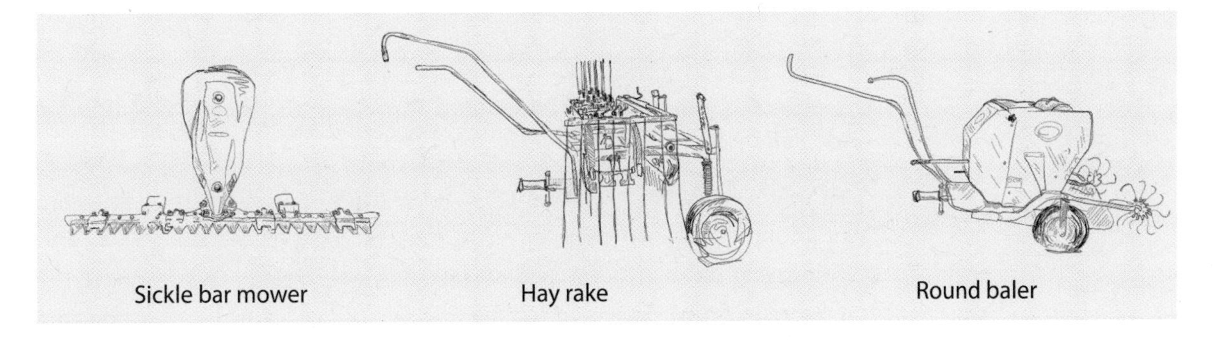

Pro Tip: *Rotation of fields between hay and pasture balances fertility and reduces weed pressure (like thistles) that grow rampant in fields that are only pastured. With this rotation, the sickle bar mower cuts the young thistles before they seed.*

GROUNDSKEEPER EQUIPMENT GUILD

Most equipment here could serve a **Landscaper** who installs gardens, orchards, and lawns. The **Groundskeeper**, however, mainly maintains these features and may need more equipment for a *full complement of lawn care services.*

- A **Dethatcher** removes layers of thatch and improves the aeration of a lawn, which helps to fill in dead patches. This implement has a 26" working width.

- A **Compost Spreader** applies compost to improve lawn growth.
- A **Lawn Mower** maintains a lawn at a precise height and can either side-discharge or bag the clippings.

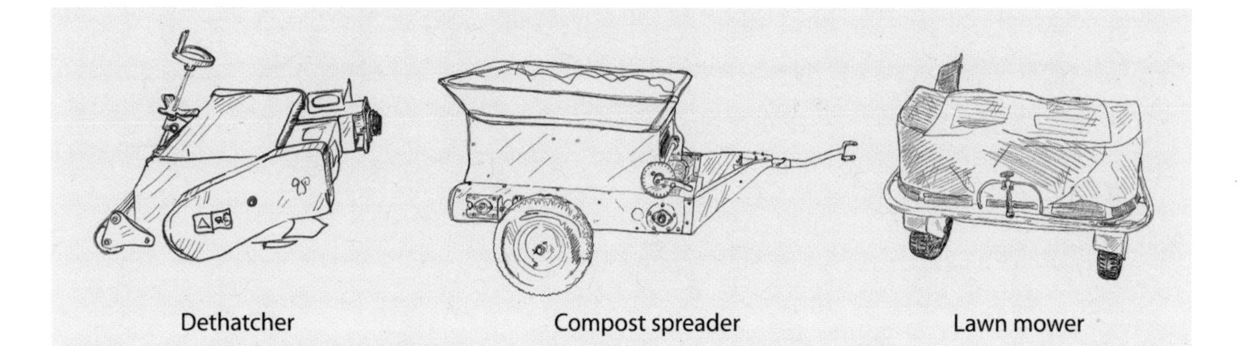

A tractor with a differential or hydrostatic drive is essential for groundskeepers because you can change speeds seamlessly if you need to slow down around a turn or in thicker grass stands. You can also do finer mowing of lawns by mowing in forward and then reversing simply shifting by moving the hydrostatic control lever (see #20 on page 27) from forward to reverse.

***Pro Tip:** Bagged grass clippings are a great nitrogen-rich amendment for compost at the base of fruit trees. This is a key advantage of a lawn mower implement for Permaculture and home orchard management.*

HAND TOOL EQUIPMENT GUILDS

Hand tools are popular for many enterprises. Whether to use them—or which to use—depends on the grower's scale and preferences. Hand Tool Guilds should be considered for successful start-up, alongside two-wheel equipment.

***Pro Tip:** Always consider how two-wheel tractors complement other scale-suitable equipment.*

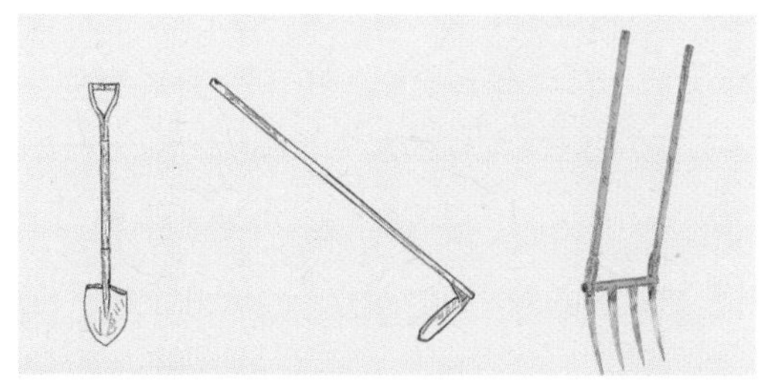

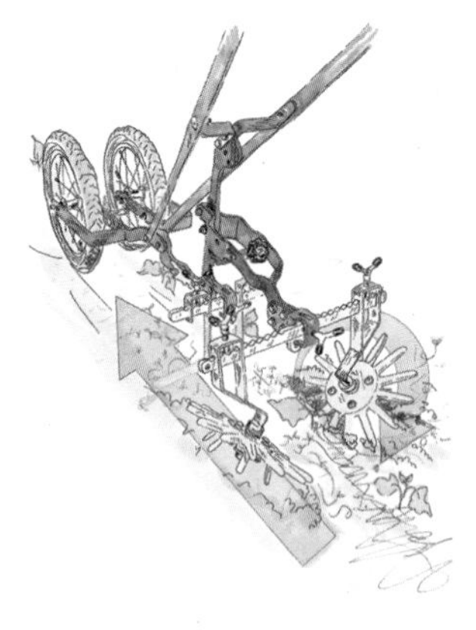

The Bed Builder Tool Guild: ergonomic shovel, grub hoe, and a broadfork are great start-up tools. This wheel hoe from Terrateck is well-suited to complement cultivation equipment for professional growers at scale-up.

DESIGN BOX
SPECIALIZED JOBS OR COMPLETE OPERATION CYCLES

Two-wheel tractors be can used with a single implement to perform **specialized jobs** (like greenhouse bed preparation), and sometimes other equipment is used for larger acreage (including 4-wheel tractors). Two-wheel tractors can also employ an equipment guild to complete entire **operation cycles** (plowing a field, forming raised beds, and finishing seedbeds). They can even be used to manage the productivity of **guild enterprise productions**, which are a companionship of different enterprises under one management, such as a market garden, orchard, or honey production. Here, more equipment works together to tackle a variety of dissimilar tasks, but sometimes the same equipment can be multi-functional, such as chipping wood for garden paths from orchard tree debris or pumping water for irrigation as well as refilling bee watering stations.

Left and top center: *The ambitious growers at Cascina Fraschina operate a large market garden in Lombardy, Italy, that produces rice, summer vegetables, and low-tunnel greens. Their two-wheel tractors are used specifically for low-tunnel bed preparation, and four-wheel tractors manage larger fields.* Bottom center and right: *On her farm in the Dolomite region of Northern Italy, Francesca grows a plethora of organic vegetables, and her two-wheel tractors complete the entire operation cycle for all land under production, including greenhouse crops and field crops, such as heirloom corn and summer vegetables.*

GUILD ENTERPRISE PRODUCTION

When deciding to grow food, we want everything! Focus is the name of the success game, while diversity is the name of the resilience game ... so, how do we balance these? **Guild enterprise production**[2] is about *focused diversity* for your land-based business. Let's look at how this model helps orient grower goals and equipment choices over scale phases.

Growers should work toward a harmony of no more than three enterprises that *balance, share,* and *inform* one another's land-use, seasonal labor, equipment, and other scale principles (refer back to Figure 1). Sometimes, enterprises are added as you scale-up; sometimes all enterprises are initiated at start-up and scale-up together.

Goals for intended enterprises and relationships between enterprises, such as equipment sharing, should be made from the get-go. Organize your intended enterprises around access to land, your vision, skills, labor availability, and management style. Then use your enterprise goals to bring in the right equipment as you scale-up to avoid pitfalls and costly mistakes.

ENTERPRISE TYPES

There are many enterprise types and many different ways of organizing them. First, we have **DIY** vs **professional enterprises**. Are you producing to save money or make money? Next, a guild of enterprises can be either more **similar** or **dissimilar**. Examples of similar enterprises include a DIY grower specializing in spring, summer, and fall gardening, and a professional grower producing summer CSA crops, wholesale garlic, and root cellar vegetables. Also, enterprises often can be defined as more **intensive** or **extensive**. A backyard garden production with three enterprises becomes a Back-to-the-Lander at a certain point, where the acreage under production justifies a new term. A Market Grower could turn into a Row Crop Farmer, etc.

Remember, aim for three, and then if you want to specialize further, consider doing this within your enterprises. For instance, the Ecosystem Grower (example below) could add different laying hens to have a kaleidoscope of egg colors on offer, or add new fruits to their orchard. Meanwhile, adding beekeeping to their farm would be exceeding the recommended number

2. This model is detailed in my book, *The Permaculture Market Garden.*

of three enterprises. In this case, they should consider having someone else steward bees on their land!

FIGURE 12: GUILD ENTERPRISES

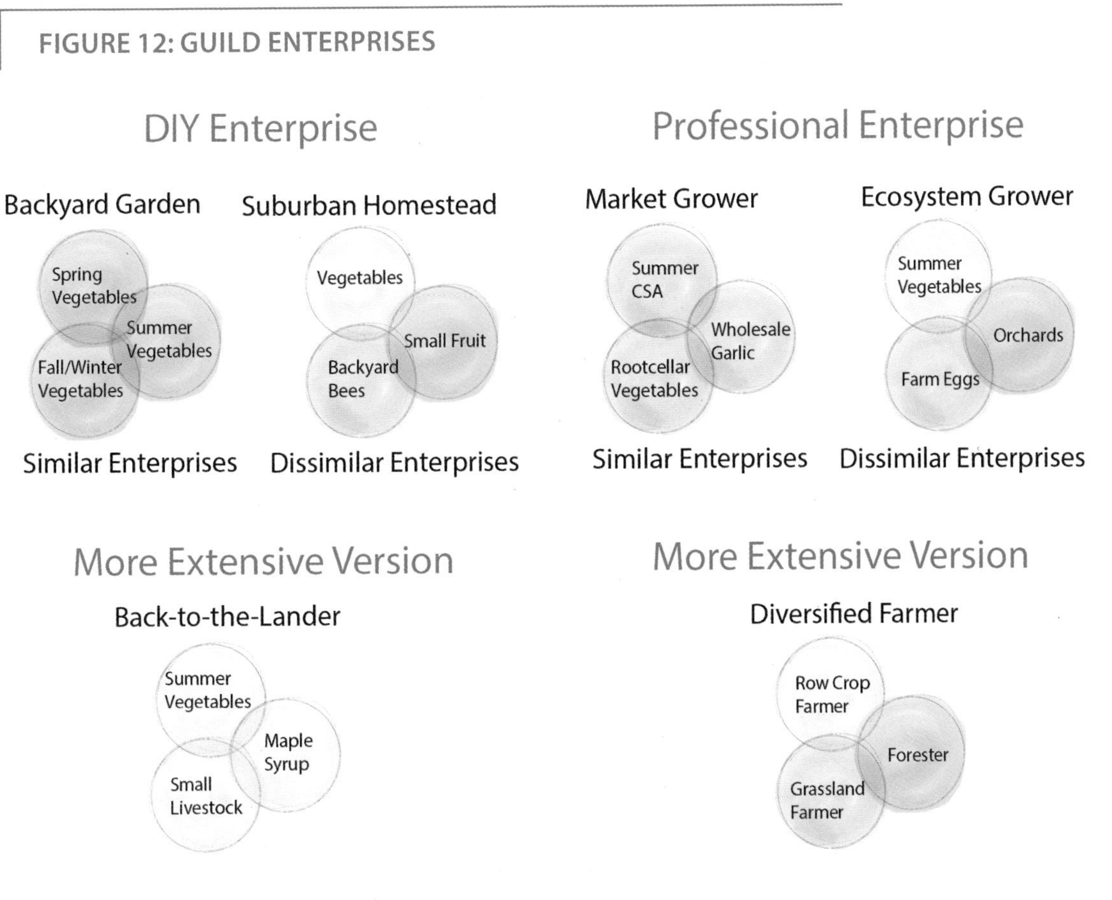

These are examples of some different types of enterprises for guild enterprise production.

ENTERPRISE SIMILARITY

Dissimilar enterprises grow many food types (animal, annual, perennial, fungi, etc.). *Think of this as diversity at the biological kingdom of life level!* **Similar enterprises** focus on a production type, such as annual vegetables, and are *still* diversified, but at the *species* and *variety* level.

Marco grows a diversity of broccolis as a distinct enterprise. His similar enterprises include mixed summer vegetables, greenhouse crops, and specialty brassicas.

SCALING-UP ENTERPRISES

WAYS TO SCALE-UP ENTERPRISES INCLUDE:

- **Expansion of a DIY-type enterprise** to become more productive. **Example:** A backyard garden becoming a homestead and finally expanding into a Back-to-the-Lander with more or less similar DIY enterprises.
- **DIY-type enterprise becoming a professional** enterprise with more productivity and commercial sales. **Example:** A backyard garden including market garden and/or other commercial productions as a guild enterprise production.
- Scaling-up can involve **expansion within your enterprise to be more specialized** and/or adding more actual production acres. **Examples:** A Market Grower scales-up production of specific crops, forms a seed garlic business, or root cellar sales are added to original summer vegetable production.
- Scaling-up can involve **expansion by adding other enterprises** that complement your start-up enterprise. **Example:** A Market Grower who starts an orchard and adds a chicken tractor to be rotated among the fruit trees or within the cover crop garden plots.
- ***Note:*** Landscapers, Groundskeepers, and Custom Farming are their own enterprises, and they can use their equipment to *professionally help others scale-up* (by tilling a lawn to become a market garden, for example). Using and providing these services is another way two-wheel tractors contribute to the new food revolution with edible landscaping and custom small farming.

Sometimes, scaling-up involves an **expansion of a DIY-type enterprise** *to become more productive.*

DESIGN BOX
EQUIPMENT FOR EVOLVING ENTERPRISES

Consider the example of a Market Grower whose intended static scale is 3 acres and a guild enterprise production with three similar enterprises: intensive growing of summer vegetables, Permabed orchard producing small berries and fruits, and extensive growing of root cellar crops, such as carrots and beets. Equipment is purchased in phases to spread out the cost and ensure proper decision-making. Initial equipment is purchased in light of eventual scaling-up. The strategy here is to:

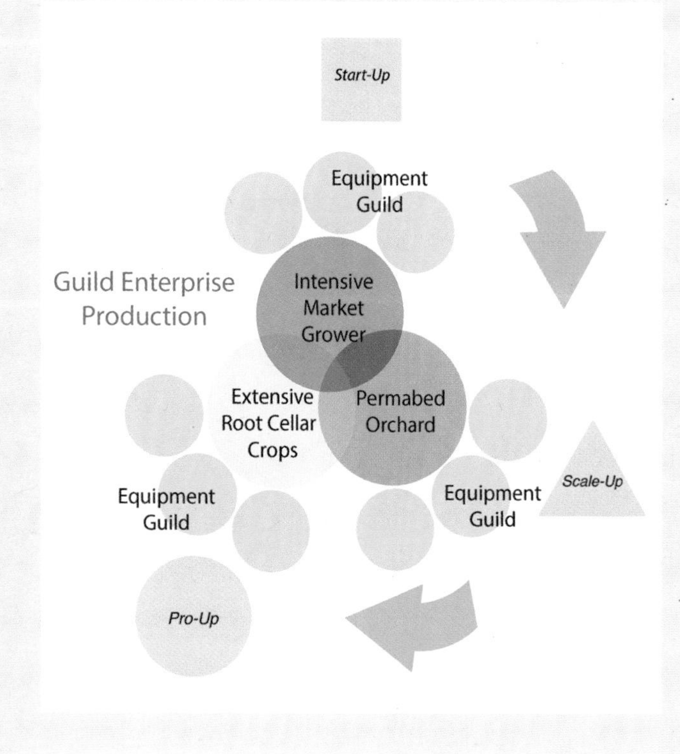

- **Focus on one enterprise at start-up,** and invest in an equipment guild that covers operation essentials.
- **Scale-up by improving your initial enterprise and begin integrating your second enterprise.** Assess how your initial equipment guild can help your second enterprise, too. Then bring in a second equipment guild that covers any essential tasks for your second enterprise that are not covered by current equipment.
- **Add your third enterprise, and choose equipment that solves weak links** across all three enterprises. In the pro-up phase, you should focus on refining systems, not necessarily adding lots of new equipment. Adding accessories or shifting strategies to maximize current equipment is the priority—then purchase for weak links.

- ***Note:*** Growers may not add enterprises in a linear sense (one in each scale phase); they may be working with all three from the get-go. The crux is focused on equipment investment in each scale phase to complete operation cycles—always considering multi-functional uses across different enterprises. Be strategic. Don't over- or under-invest!

FARM FEATURE

LUCA'S ENTERPRISES

Guild enterprise production is about a mix of enterprises that create resilience (not putting all eggs in one basket). Luca's farm at the foot of the Dolomite Mountains in Northern Italy grows a variety of field row crops, such as melons, brassicas, and radicchios, alongside specialized greens and tomatoes in low tunnels and a fruit orchard with apricots, persimmons, and figs. He uses WhatsApp to manage orders for his CSA basket program. His two-wheel tractor is over 30 years old and still kicking!

FIGURE 13: SCALING-UP TO GUILD ENTERPRISE PRODUCTION

Sometimes growers will scale-up by evolving their production from DIY to professional.

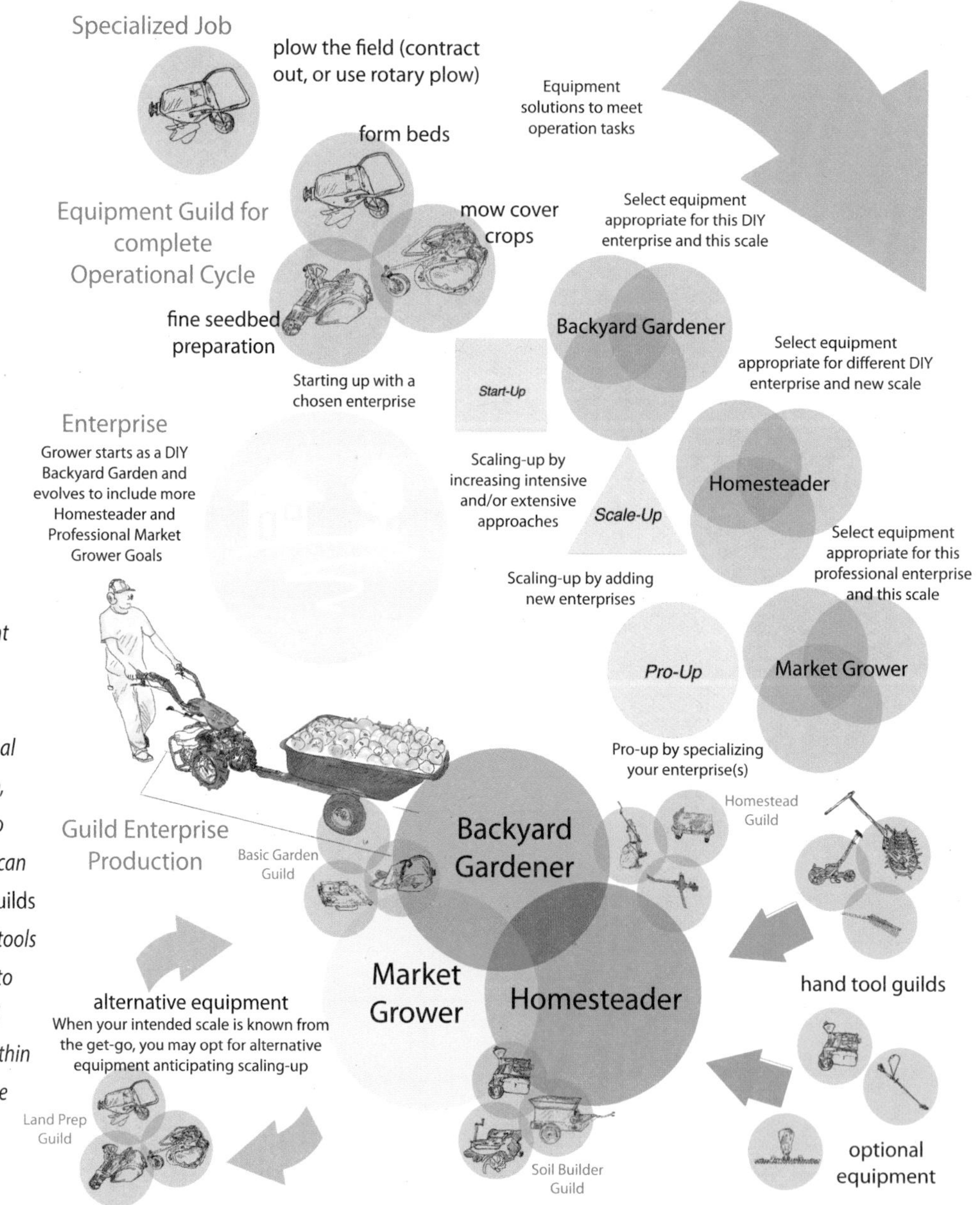

There are equipment solutions for any specialized job or complete operational cycle. From start-up, through scale-up, to pro-up, equipment can be understood as guilds of implements and tools that allow growers to meet their intended production scale within their guild enterprise business model.

DESIGN BOX

ZACH'S FARM EVOLUTION

Farms evolve over time, so plans for equipment change, too. My own project started as a backyard garden and grew to be a market garden. My guild enterprise production included: **1. vegetables and fruits, 2. heirloom seed garlic,** and **3. education/design,** and then it transitioned its focus when I started my design business and nursery.

MY GUILD ENTERPRISE PRODUCTION LOOKS LIKE THIS TODAY:

1. **Heirloom and open-pollinated plant/seed nursery.** We grow many varieties of garlic, trees, berries, herbs, and more. Trees sold are *only* those we have experimented with in situ in our climate and region and can offer real advice for! We sell fruit, but mostly we keep seed/scions for propagation.
2. **Edible Ecosystem Designer/Installer** for small farms, communities, and homesteads.
3. **Research and Education.** In partnership with like-minded groups and The Ecosystem Solution Institute, we conduct research and create eco-education sites. Our flagship site is The Edible Biodiversity Conservation Area (EBCA), a 100-acre heirloom and open-pollinated edible ecosystem laboratory.

MY EQUIPMENT ADAPTATION

Equipment once used for the production of vegetables for hundreds of CSA shares and market customers now maintains the nursery plots and helps with landscape installs. A few two-wheel tractors and a few dozen implements and tools fit in my 6 ft ×12 ft cargo trailer, with my truck bed full of plants. Well-equipped, we can remove weedy vegetation, micro-plow fields, form and finish Permabeds, plant trees and vegetables, and ultimately completely transition land to a custom diversified edible ecosystem! A similar scale of landscape work with single-purpose equipment or four-wheel tractors would require a flatbed trailer, large trucks, and much more money invested.

GUILD ENTERPRISE BRAINSTORMING TEMPLATE

Understanding your guild enterprise production is critical for farm planning, and it is never too late to start! This template can be used to help brainstorm your guild enterprise production by asking key questions and providing space for equipment guild selection in a balanced manner. **You can request blanks from ecosystemU@gmail.com.**

FIGURE 14: ENTERPRISE BRAINSTORM TEMPLATE

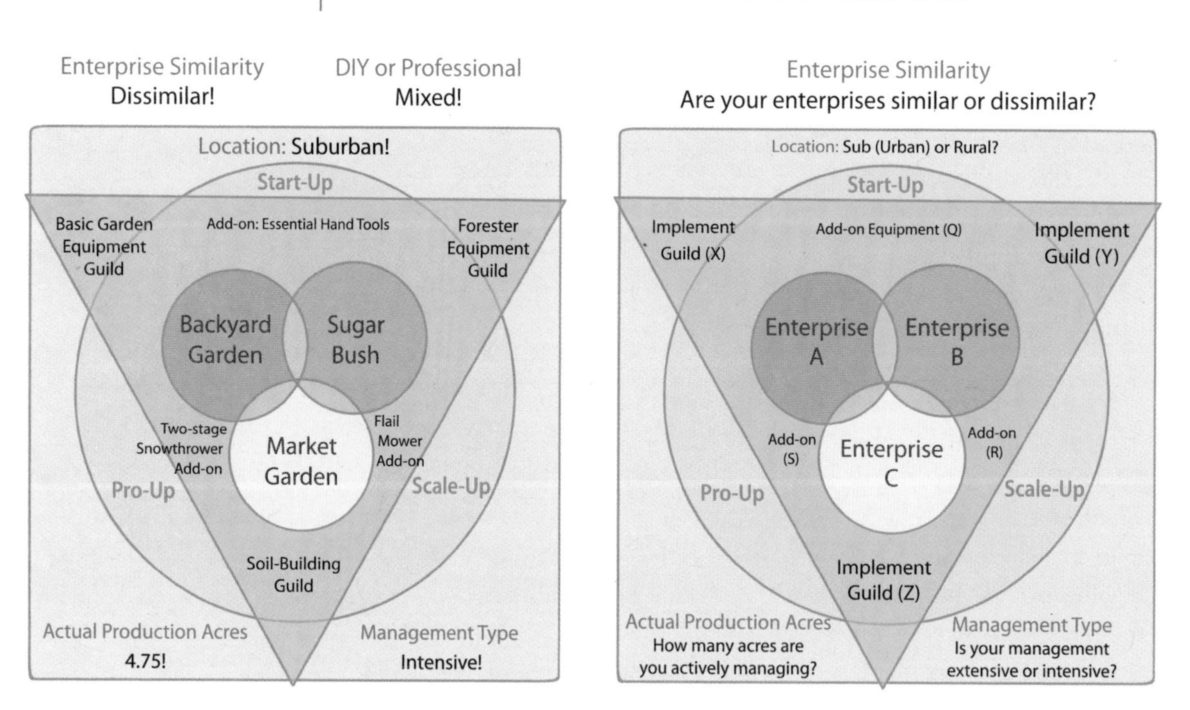

5-ACRE PROPERTY EXAMPLE

This is an example for a 5-acre property using the **Enterprise Brainstorm Template.** See key questions below with example answers. Use these for your own brainstorming.

EQUIPMENT DESIGN TESTING QUESTIONS

1. *Are you suburban, urban, or rural?* **Example:** Suburban, just within city limits, zoned rural residential.
2. *Are you extensive or intensive?* **Example:** Intensive, often 3 or 5 rows per bed, 12" footpaths, 2–3 successions per bed per season.
3. *What are your actual production acres?* **Example:** 5 acres, 1.75 arable, .25 for buildings, and 3 in woodlot.
4. *What are your intended actual production acres?* **Example:** 4.75 acres will be used actively for production; no other land is going to be rented or used.
5. *What is your intended guild of enterprises?* **Example:** Backyard Garden/ Sugar Bush/Market Garden.
6. *Are these enterprises similar or dissimilar?* **Example:** Dissimilar; I will need to invest in more types of equipment, skill-building, etc.
7. *Are enterprises DIY or professional?* **Example:** Mixed; this model will provide both profit and enhance well-being and resilience for me and my community.
8. *What is your scale-up timeline for different enterprises?* **Example:** First Garden, then Sugar Bush, then Market Garden over 3–4 years.
9. *Will you mechanize most operations as you scale-up? Why?* **Example:** Yes, because I have a shortage of consistent labor, and their time will be focused on harvest and other tasks that cannot be tractor-mechanized easily. Hand tools will complement the basic garden guild at start-up and will continue to be used, but implements will replace some key functions later—like opting for a compost spreader and utility cart instead of a wheelbarrow and extra hands for compost jobs and hauling.
10. *Which equipment guilds will accomplish tasks and operation cycles for each scale phase?* **Example:** Basic Garden, Forester, and Homestead guilds.
11. *Which equipment guilds will accomplish all needed operation cycles for your static scale?* **Hint:** These *should* be the same as in number 10! **Example:** Basic Garden, Forester, and Homestead guilds.

EXAMPLE OF 5-ACRE SCALE PHASES

Start-up with the **Basic Garden Guild** for **backyard garden enterprise**. Rent or hire out plowing, then use the *hiller/furrower* to form Permabeds, and the *buddy cart* to move compost, supplies, and harvests. **Option:** Choose the **Land Prep Guild** at start-up if you are sure about an intended market garden at pro-up. Ask yourself: What is my goal?

Scaling-up with **Forester Guild** and adding **maple syrup enterprise**. A *wood splitter* processes undesired trees into firewood for evaporators, and a *utility trailer* is used to move firewood and syrup collection pails to the sugar shack. A *shredder/chipper* makes use of excess branches as chip mulch for trails in woodlots and paths in a garden—carted around using the *buddy cart*. So, we see how enterprises and equipment work together! A flail mower is purchased as an add-on to help with garden crop debris management and Sugar Bush path mowing.

Pro-up with the **Soil-Building Guild**. This adds a compost spreader, power harrow, and power ridger. The *spreader* is a welcome relief for compost management, and it allows seasonal applications of compost made from woodlot leaf fall (windrowed and turned with a tiller). The *power harrow* improves soil structure with better depth control and no mixing of layers. The *power ridger* is used to reform beds and power compost the paths. **Add-on:** A two-stage snow blower can be purchased to improve overall property access in winter and early spring.

Note: Equipment is never obsolete and always multi-functional. The buddy cart has continued use for hauling vegetables and for odd-jobs around the Sugar Bush lot. The utility trailer is employed to move transplant trays from the new greenhouse to the fields.

Pro Tip: *It's important to understand your property's constraints (like the soil type and current land use), set intentions for scale (actual acreage), make goals for enterprises (garden, syrup, market garden), and then select equipment guilds and organize their purchase for different scale phases.*

BUDGETING FOR TRACTORS AND EQUIPMENT

Part of the equation for equipment decision-making is literally the *equation—the mathematical one.* We need funds to purchase equipment, and we need to use those funds wisely. So taking the time to do the math before spending is quite important. Part of this is understanding how equipment saves money as well as how it makes it, and how the way you manage your land with equipment can lead to profit—but also resilience in the face of socio-economic and environmental change.

PROFIT RESILIENCE PRINCIPLES

THE GOAL IS PROFITABILITY AND CASH FLOW

- The goal is profitability, whether for a DIYer or a professional. A Homesteader should have a return on investment in *savings* as much as a commercial grower makes *income.*
- Profit is a measure of income *minus* expenses. This should also include a wage for the farm owner(s). So, profit should be what is left over after all expenses are met, including a reasonable hourly wage.
- Profit can then be invested back into the business or divided as dividends among the farm owners or business co-op members.
- Profitable farms can afford to invest in themselves. However, profit *isn't* a reason to buy equipment! Rather, equipment decision-making, as we have seen, comes from understanding scale and creating goals for a static scale of specified enterprises.
- Yet, profit does contribute to *cash flow*, alongside grants, loans, and prior savings.
- Enterprises can be seen as either shorter- (vegetables) or longer-term (orchard) profit centers.
- Investment in equipment for short-term enterprises can be readily justified with annual earnings. Long-term profit centers should rely on grants and sound budget-backed loans, as real profit takes more years.
- Crowd-funding and advance support from customers is another source of start-up cash flow. Feeding your community and gaining support is

one reason to grow diverse foods that provides food security *and not just* high profit per square foot production!

PROFIT PER SQUARE FOOT AND PRODUCTION RESILIENCE

- Profit per square foot/acre is a *holy grail* for Market Growers. However, this term should be taken with a grain of salt because profitability is really *much more than* income and expense. Profit measurement should also include *other* forms of farm production and ecosystem services. For instance, a highly diversified farm can produce and hold soil organic matter, which has real value in fertility savings, soil biology inoculation, and conservation, and, thus, enhanced yields and long-term resilience.
- Also, sometimes intensive/extensive management will affect profit per square foot relative to farm resilience. An intensive market garden on a ½ acre that grows *only high-value lettuces* and is highly specialized can be very profitable per square foot—even more so for greenhouse growers. However, such farms can be reliant on imports for all inputs: seed, compost, mulch, supplies, etc. This isn't resilient farming … but *it is very profitable*, and we do need this model of production!

INVESTMENT IN EQUIPMENT FOR RESILIENCE

- **More extensive farms might use agroecological methods** like green manures, intercropped perennials/annuals, and in situ mulch production of straw and ramial chipped wood (RCW). This will require more equipment investment—justified not in *higher profit* per square foot, but in more *profit resilience*! Farms that control the quality of their weed-free compost and mulches can save on imported mulch costs; these savings can be invested in other farm resources, such as ponds or long-term tree crops. The testimony to a lack of investment in resilience comes with droughts, seed shortages, and poor-quality mulches blooming scutch grass in your garlic!
- **Farms growing mixed vegetables often have less profit than lettuce farms** because they grow lower profit/square foot crops. They also will often require more tools and equipment to meet their diverse needs. Again, this equipment investment is justified in social-capital building by providing a well-rounded diet and food security to your community that can support you in hard seasons.

- **Farms investing in ecosystem design** are literally growing future profit potential. When trees and crops are layered, there is actually more net primary productivity; more sunlight is captured and converted to crop and fruit. Examples of ecosystem farming include intercropped Asian pears, hazelnuts, and black raspberries, with intensive lettuces thriving in the dappled shade in August and rows of self-seeding kales that are ready to bunch before you can even prepare a new bed top! Here, equipment to manage perennials is justified by the long-term profit potential.

Pro Tip: *Diversified farms with both short- and long-term production are not only profitable and financially resilient but are often operating in a way that respects our Earth's finite resources and helps create a resilient society. The integration of annuals and perennials protects heirloom cultivars, builds habitat for endangered species, mitigates climate change, and fosters community food security, safety, and sovereignty.*

SO, WHICH EQUIPMENT TO INVEST IN?

- The ½-acre start-up farm focused on only high-profit crops has fewer equipment needs and will focus on very specific equipment guilds for operation cycles. They will often include very specialized equipment for harvest and post-harvest handling in the scale-up and pro-up phases.
- On the other hand, the more extensive and diversified farm will need more equipment types. They will often add new enterprises over the years and bring in equipment guilds to suit. The key is always considering how your implements can serve different functions across your multiple enterprises to keep costs low.

WORD TO THE WISE

- As we discussed, profit can be deceiving as a benchmark for when to invest in equipment.
- Growers can make $100,000 on 1-acre, or $300,000 on 2-acres, or $400,000 on 5 acres. Remember, (**income – expense = profit**). Increasing income *or* decreasing expenses can increase profit.
- Sometimes, equipment is needed because it produces a high profit per acre, but equipment may also be valued by input savings, social capital building, long-term profit aims, and overall resilience.

Paying attention to these factors can create high net profit in a single growing season—as much as $1,000 to $3,000 per 100 ft bed ($100,000 to $300,000 per acre).

FIGURE 15: HIGHEST PROFIT MATRIX

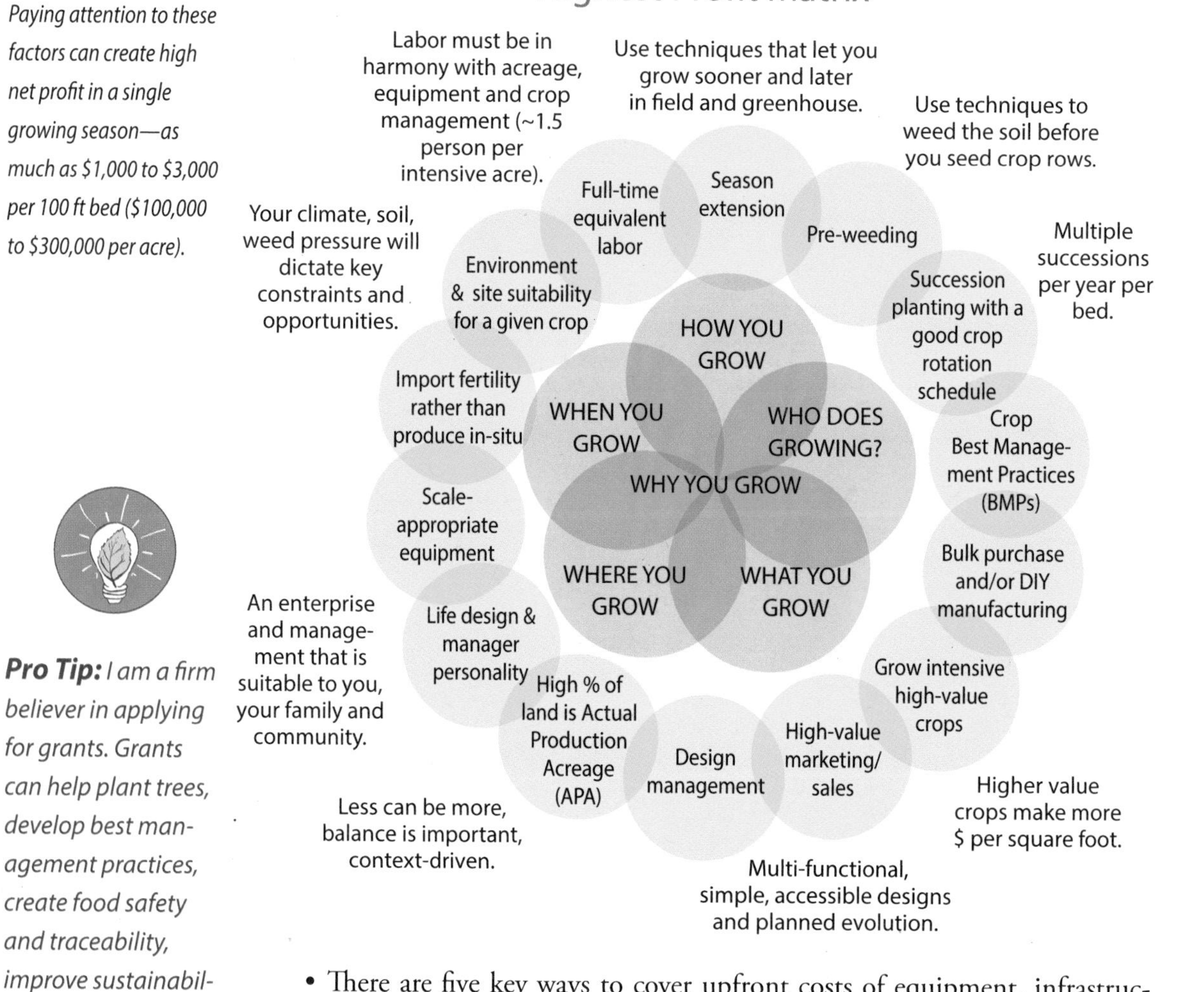

Pro Tip: *I am a firm believer in applying for grants. Grants can help plant trees, develop best management practices, create food safety and traceability, improve sustainability, create green jobs and train youth, etc. You can fund your start-up with grants and continue to fund your scale-up and pro-up too!*

- There are five key ways to cover upfront costs of equipment, infrastructure, and initial supply purchases: 1) use savings, 2) get a business and/or family loan, 3) get project-specific and other grants, 4) use an upfront income model like Community Supported Agriculture, or a Kickstarter-type scenario, and 5) maximize government and other program start-up funding. Starting my own farm, I used most of these over the first five years as I scaled-up to hundreds of CSA shares, two large urban farmers markets, on-farm events, and online sales.

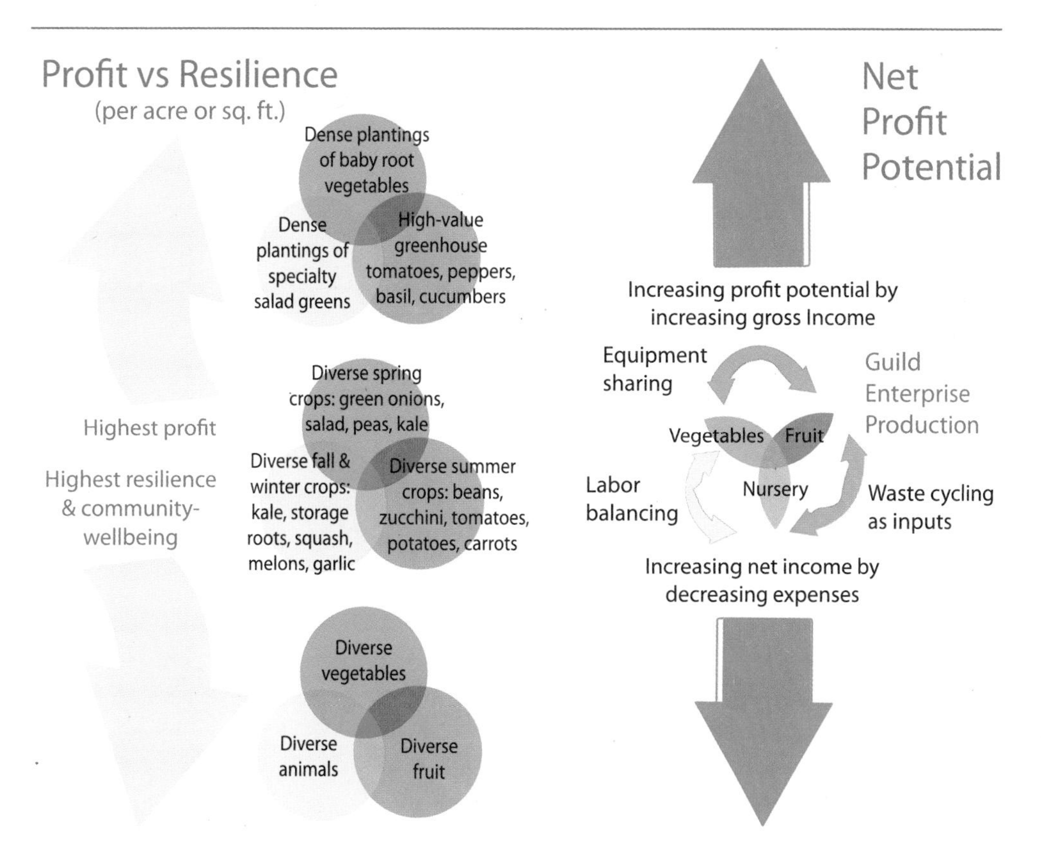

EQUIPMENT PURCHASE SOUNDING BOARD

Now, let's go through a series of questions to help sort out which tractors and equipment are best for different growers.

WHICH TRACTOR OPTIONS SUIT MY ENTERPRISE(S)?

Professionals prefer models with features suited to specific operation cycles, such as M2w-tractors or row crop tractors with specific implements (sometimes both are needed). Smaller DIY enterprises often find that tractors with lower-cost basic features are sufficient for part-time use. For gardeners and homesteaders, the M2w-tractor is all that is needed.

Backyard Gardener Example: A smaller horsepower tractor with fewer features is perfectly capable of preparing your garden each year and doing other jobs around the property, like snow blowing. ***Market Grower Example:*** A tractor with a differential drive can make maneuvering between plots much easier, and it also allows you to lock the differential for field tillage. Also, choose a model that has enough energy for the power-hungry flail mower. ***Grassland Farmer Example:*** A hydrostatic drive would be well-suited to operations with a lot of mowing.

Which implements can be of use to fulfill my specific job?
Sometimes a single implement is all you need—or it may be an important add-on that is needed for a specific job.

Backyard Gardener Example: *With a ¼-acre garden, a rear-tine tiller with a furrower could be all you need.* ***Nursery Example:*** *The power ridger can both furrow for planting trees and fill in those furrows. If trees are all you grow, then this may be all you need!*

Which implements work well together to complete my seasonal operations?
Professionals opt for equipment to finish entire operation cycles in a timely and profitable manner. DIY enterprises tend to use equipment that meets several needs and is more likely to be of use through all four seasons. However, the same selection process should be used for both types of managers: group implements into guilds that can complete operation cycles when possible!

Market Grower Example: *Bed preparation is a key operation cycle that can be mechanized. The rotary plow forms new beds, the rear-tine tiller with PDR makes a seedbed, and the flail mower mows cover crops. These three implements can be used to form an equipment guild for land preparation. This Land Prep Equipment Guild is popular with many Market Gardeners.*

Can your equipment guild be used for other operation cycles?
Market Grower Example: *The Land Prep Guild is multi-functional! The flail mower can mow an initial stand of cover crops or weeds when clearing land; the rotary plow can plow new land into furrows for a future garden; and the rear-tine tiller can be used to break up the furrows for future bed forming.*

Can your enterprise use more than one set of equipment?
Yes, as you scale-up your operation, you will often find new equipment can be brought on as a guild.

Market Grower Example: *At start-up, the Land Prep Guild is great, while scaling-up with a Soil-Building Guild is a good choice.*

What is a reasonable budget for a two-wheel tractor and equipment?
As a wise farmer once said: "It is easy to buy too much heavy metal too quickly!" By assembling equipment into ***guilds*** *and having an* ***intended scale,*** *you can budget accordingly. A two-wheel tractor starts at around $2,000, and most professional growers will spend around $4,500. Three implements will usually cost around $4,000, and a second equipment guild to scale-up might add an additional $3,000.*

A Marker Grower *with less than 1.5 acres is looking at a two-wheel tractor and equipment budget of about $8,500. See a full breakdown in Figure 16.*

BUDGETING FOR MARKET GARDEN SCALE-UP

When budgeting for equipment guilds, consider your static scale. For this budget example, we will follow the equipment options and costs for a Backyard Gardener (¼ to ⅕ acre) at start-up who scales-up to become a Market Grower (1.5-acre) and then includes some extensive row crop production of bulk root vegetables (carrots, for example) in pro-up. Additional expenses, like tools, other equipment (wash station setup), and capital investments (land, trucks), should always be considered alongside two-wheel tractor investments in any budget.

FIGURE 16: BUDGETING FOR EQUIPMENT GUILDS

The table shows a series of tractor and equipment budgets for enterprises that are related. The table can be understood as showing three distinct enterprises, or it can be viewed as showing *a process of scaling-up from one to the next*. These equipment budgets and income estimates are based on highly profitable enterprise management. Know-how is also important, and this is why a Backyard Gardener will never make as much in sales or savings as a professional Market Grower on the same acreage. Skill-building and proper equipment/supplies are required for bigger sales/savings. Local markets are also important to consider; in some areas, prices for fresh local organic produce can be double the price of grocery store produce.

BCS 739 with electric start	Rotary Plow	Tiller (30") + PDR	Flail Mower (30")	Power Ridger	Spreader + expansion kit	Power Harrow	Total
$4,875	$1,725	$995 +$915	$2,725	$1,875	$2,425 + $310	$2,615	$18,460 New $12,434 Used

Note: Pricing is in USD and is current as of 5/2022.

CHAPTER 5

Selected Enterprise Examples

Let's explore a number of DIY and professional enterprises and consider suitable equipment, scale phases, and operation cycles. This will reveal how the same equipment guild can apply to different enterprises to accomplish various operation cycles.

Enterprises have been given common names, such as Homesteader, Market Grower, Grassland Farmer, etc. These names are meant to help readers more quickly identify their type and review equipment complements. Growers will see how different types can form a guild enterprise production, maximizing beneficial relationships between enterprises.

DESIGN BOX

UNDERSTANDING TILLAGE AND CULTIVATION

The terms *tillage* and *cultivation* are used interchangeably. But having some clearer definitions is helpful.

UNDERSTANDING TILLAGE

The goal of tilling is usually to prepare ground so it is loose and ready to be planted or seeded by hand or machine. When you come to an old hay field or lawn, you need to turn this ground into garden soil. The most common practice of doing this involves tillage. Sometimes tillage is given a narrow definition meaning "worked by an actual tiller" (the implement that mixes the soil), and sometimes it is defined more broadly as the process of working the earth with any number of implements: rear-tine tiller, power harrow, or any other similar earth-working tool. I like to speak about tillage broadly, letting the term refer to a number of popular ways we work the soil.

TYPES OF TILLAGE

1. **Primary tillage** (often 5–8" deep) includes plowing, discing, and preliminary tilling for land turnover.
2. **Secondary tillage** (often 3–6" deep) involves forming garden beds and initial preparation

of the bed top. Implements used might include a rotary plow and rear-tine tiller or a power harrow.

3. **Seedbed preparation** (often 1–6" deep) uses a power harrow or PDR tiller for seedbed preparation. The tilther tool from Johnny's Selected Seeds can be used *in light soil conditions*. A simple rake can be used to refresh the soil surface for seeding. Multiple passes with a tine rake or basket weeder for pre-weeding are done at this stage.
4. **Cultivation** (.25–3" deep) is the lightest tillage, used for weeding and improving soil surface. Hilling cultivation is deeper (2-6"); soil is mounded up around the stalks of crops (like potatoes) for support and improving growth. Also used to manage weeds.
5. **Reforming** (~3–5" deep) and **chisel plowing** (~12" deep) can improve soil structure. Use a rotary plow, furrower, or power ridger to move path material back onto bed top. Chiseling breaks up compaction layers and improves drainage in the early years.

UNDERSTANDING CULTIVATION

Cultivation should be understood as a *series of field passes* that keeps a crop free of between- and in-row weeds. Cultivation equipment consists of toolbars with different cultivators *assembled differently* for *specific* row crop management. This often entails one cultivator following another to improve the overlap of cultivation so you can achieve in- and between-row cultivation in the same pass.

TYPES OF CULTIVATION

1. **Tilthing** is for shallow preparation (as opposed to *tilling*) of an entire bed top before seeding to improve *soil-to-seed* contact. Using a tine weeder or basket weeder is preferred, though using a tiller or power harrow at very shallow settings can produce good results.
2. **Pre-weeding** is light cultivation of the entire soil surface to improve a seedbed before crops are planted. A tine weeder or basket weeder is often used.
3. **Blind weeding** is for light, full-bed weeding of white thread-stage weeds while crops are growing. A tine weeder is best for this job.
4. **Between-row weeding** is done in single- or multiple-row setups for weeding between the crop rows. Various cultivators are commonly used and available: *duck-foot, beet knives,* and *tender plant hoes*.
5. **In-row weeding** is done with sweeps to remove weeds. Work the soil lightly (cultivate) within the actual crop row, e.g., between the actual onion plants. This often entails finger weeders or mini-discs for micro-hilling, the former physically disturbs the soil between crop plants, while the latter acts to cover over the smaller weeds in the row without covering the larger crop plants. Some equipment, like **torsion weeders**, provide in- and

between-row weeding; these weeders are making a comeback as great cultivation tools.

6. **Pathway** cultivation is done between beds or rows with a gang of S-tine, C-tine, or chevron (*duck-foot*) cultivators. Market Growers usually have actual paths between wider raised beds. For Row Crop Farmers, the spaces between all rows in a field are the same, and path cultivation is done by a tractor driving between wider-rowed crops rather than straddling them for cultivation.
7. **Disc hilling and spyder hilling** are forms of in- and between-row weeding. **Disc hillers** are for more aggressive weeding and hilling actions, and **spyders** serve a similar purpose, but they provide more aeration, are less aggressive, and leave a rougher surface (which is less erosion-prone). **Bio-discs** are a less aggressive version for small crops. ***Note:*** Hilling is only used for the type of crop that isn't bothered by having its stem partially buried, such as potatoes, beans, and corn.
8. **Dust mulching** should also be mentioned here, as it is the result of cultivation. Once the soil's surface crust is lightly broken, the cultivated soil settles on the surface in layers of dust, like a mulch. This prevents further water loss from the soil because the cultivation has interrupted the capillary action in the soil between the A horizons and the air above.
9. **Deep cultivation** is just what it sounds like. It can be done by running a gang of C- or S-tines through the soil as part of land preparation or re-preparation to remove aggressive weeds, like grasses. This requires an M2w-tractor.
10. **Edge-of-bed cultivation** is done to remove weeds growing along the sides of raised beds or the edges of poly mulch.

Note: Some enterprises mechanize cultivation, and others maintain hand tools for all cultivation tasks.

THE BACKYARD GARDENER (.25–1.5 APA[1])

The **Backyard Gardener** is a serious grower in rural or semi-urban areas who is maximizing their yard for annual vegetable production but is less concerned about other property management. Start-ups on less than .5 acres will entail a mix of hand tools. Scaling-up will require more equipment. A garden of .5 to 1 acres may justify a two-wheel tractor for bed management. The one-time job of *opening new land* should be outsourced, as *that* equipment is soon obsolete at this scale. At a maximum, there are Backyard

1 Actual Production Acres=APA

Gardeners operating on 1.5 acres—primarily for home and community use; these enterprises will be well served by a two-wheel tractor and various implements.

TRACTOR RECOMMENDATIONS

A lower-cost model suitable for part-time use is commonly employed. A straight-axle and smaller-engine tractor like the BCS 722 (or the 728, in Europe) is likely to fit the bill. Larger models with extra features won't get enough use to justify the cost for most Backyard Gardeners. Tractors like the 722 have two working speeds and an additional third road speed for efficient movement from job to job. Smaller tractors are better for small properties where growers want to work close to buildings and other common lines. **Common lines** are those lines, such as fence lines and driveways, that can be used as reference points for squaring a new garden plot or laying out a new orchard row or edible hedge. These are discussed further in Chapter 7, "Garden Plot Layout."

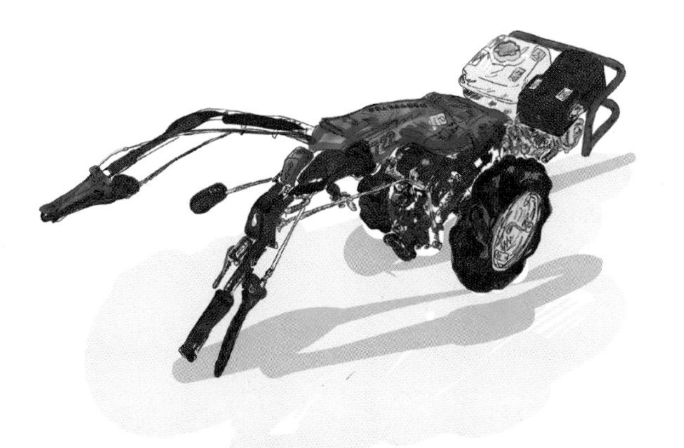

*The **722 Harvester** is a great little tractor. It has an 8-hp Honda GX240 engine, two forward and two reverse modes, and one road speed. It is very useful for preparing beds following common lines of a property.*

GARDEN LAYOUT

Backyard Gardeners will benefit from a simple Permabed layout organized along the **common lines** of their property's buildings, fences, and driveways. Imagine working from your house outward into the yard, leaving a 6 ft alley in lawn, with the rest turned into Permabeds (4 feet wide) and running the full yard's length (minus perimeter property-line alleys). A good layout allows Backyard Gardeners to maximize two-wheel tractors to make passes in the garden, mow the perimeter, and cultivate with minimum "edges."

EQUIPMENT GUILDS FOR SCALE PHASES

- **Hand tools:** Relevant tool guilds are needed for **start-up** for essential land and bed preparation, seeding and weeding. Most backyard gardeners will begin by maximizing professional hand tools before they consider two-wheel tractors.
- **Basic Garden Guild:** This is the **start-up** equipment guild for the Backyard Gardener who has plans to expand soon and make use of two-wheel tractors, even adding other enterprises, and it's a **scale-up** guild if they simply want a modest increase in their garden size over time and intend to continue to invest primarily in gardening, rather than adding other enterprises.
- **Add-ons:** There are some additional tools/equipment that can help the Backyard Gardener pro-up by targeting inefficiencies or weak links in the production cycle: a wheel hoe with finger weeders for refined in-row weeding, adding a PDR for their tiller, or even bringing in a water transfer pump for irrigation.
- ***Note:*** Backyard Gardeners that scale-up to be Homesteaders and/or pro-up to become a Market Grower will follow equipment-scale-suitability for *those* enterprises at their static scale.

SUBURBAN HOMESTEADER (1–3 APA)

The **Suburban** (or rural) **Homesteader** is a serious grower with more land management in mind beyond their gardening. Their goals are self-reliance and property maintenance with multiple DIY enterprises.

TRACTOR RECOMMENDATIONS

Homesteaders often have additional acreage to mow around gardens, orchards, and buildings, so a tractor with differential and/or hydrostatic drive is of benefit. Models like the BCS 749 (or a similar model) would fit the bill nicely and allow easy scaling-up to more commercial production if this is planned.

EQUIPMENT GUILDS FOR SCALE PHASES

- **Homestead Guild:** This is a **start-up** guild for Homesteaders who maintain multiple enterprises and need easy lawn mowing, irrigation, and even winter snow blowing within and around gardens, orchards, and pasture.
- **Hand Tools:** These are **start-up** tools that are useful to serious Homesteaders for garden beds until they want to scale-up the garden.
- **Basic Garden Guild:** This would be a good **scale-up** guild for a Homesteader who wasn't managing much garden initially and then begins to grow larger gardens later and starts mechanizing.
- **Other** equipment and tools will be acquired at **pro-up** *based on very specific homesteader needs.*
- ***Note:*** Opening new land for gardens is just a small part of *why equipment might be needed* for Homesteaders and Backyard Gardeners. Contracting this work out or renting equipment is best, allowing investment to go into other equipment.

FIGURE 17: HOMESTEAD OPERATIONS

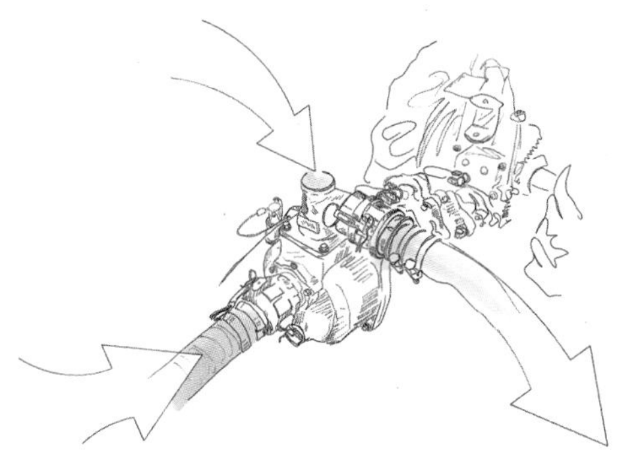

The transfer pump can be useful for small irrigation jobs. The lawn mower's collection feature helps manage clippings as a nitrogen input.

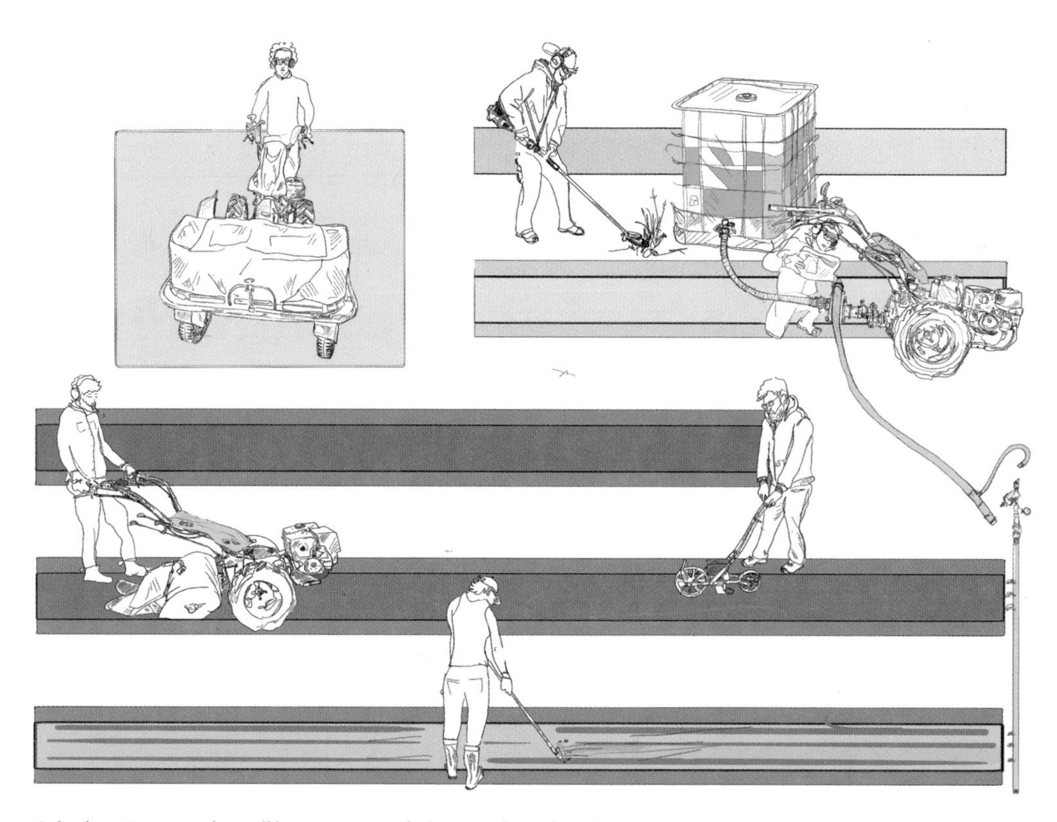

Suburban Homesteaders will have a variety of jobs to perform, from lawn mowing to irrigation, bed building and reforming, seeding, weeding, and harvest. At this scale, hand tools and two-wheel tractors are both used in the garden, and emphasis more on one or another depends on your goals. Remember to consider scale phases and bring in tools and equipment at the right time and with your intended static scale in mind.

For those with more of a perennial-oriented homestead, tillage equipment will be less useful over time. Perennial growers should focus on equipment that helps manage trees, shrubs, and groundcovers as they scale-up: irrigation for fruit trees, chippers for mulching, and flail mowers for debris and path management.

The Suburban Edible Ecosystem *in Winnipeg, run by Vera Banias, is a a full-scale suburban homestead producing annual vegetables and small fruits and doing wild edible ecosystem restoration alongside essential research on solutions for suburban land management as part of The Ecosystem Solution Institute.*

MARKET GROWER (1–3 APA)

The Market Grower always has an eye on honing equipment use. Planning for acquisition at start-up, scale-up, and pro-up is critical to make sure operations function smoothly and profitably. For market growing, a *trajectory toward a static scale* is critical. The goal is a highly productive operation that gets fresh vegetables to local markets following a clear seasonal schedule of operations with well-outlined expense budgets and income goals. Market growers are always aiming for the perfect system that allows them to be both profitable and efficient—in short growing seasons. For this reason, they are at risk of both under- and over-investment in equipment for their scale.

Understanding your markets before setting goals for scale and initiating equipment intake will be critical: wholesale accounts will justify more extensive management, whereas CSAs and farmers markets entail more intensive styles.

URBAN PRO GROWER

A variation on the Market Grower is the **Urban Pro**, who, despite working on as little as ¼ acre, is fully mechanized for very rapid bed turnover. An Urban Pro grower will often have 3 or 4 successions per bed per season; they maximize labor, use time-saving equipment (e.g., rear-tine tillers and Paperpot Transplanters), and employ season-extension techniques like row cover, weed barrier, and greenhouses.

"I still run a 739. It's the smallest size that can run all the main implements—heavy flail mower and power harrow—yet it's still lean and small."

— Curtis Stone

TRACTOR RECOMMENDATIONS

Market Growers should have a tractor that is powerful enough to do heavy earthworking. A hydrostatic drive isn't essential; choosing the proper gear and getting to work in a set speed is all that is needed for most applications. A differential drive *is* recommended, however, because it can be locked for field work and unlocked for easy turning at headlands and moving between plots. **Recommendation:** The BCS-739 with PowerSafe clutch could do the trick, or the 749 which has a larger range of implement compatibility, or the 853 with its mechanical clutch option. These are some of the most popular market garden tractors in America, with the 740 in Europe being similarly popular. A row crop tractor is sometimes used by Market Growers specializing in certain crops and employing an *extensive approach* when they have the land, but lack the labor.

EQUIPMENT GUILDS FOR MARKET GROWERS

- **Hand Tool Guilds:** Most hand tool guilds would be start-up essentials for Market Growers.
- **Land Prep Guild:** This is a start-up guild for Market Growers who rely on opening new land, seedbed preparation, and reforming new beds on a routine basis.

- **Soil-Building Guild:** This is a scale-up guild for 1–3 acres that provides added efficiency with compost and cover crops.
- **Row Seeding/Weeding Guild:** This is often a **scale-up** or **pro-up** guild for Market Growers depending on whether they focus more on transplanted crops or seeded crops. For some growers, seeding/weeding tasks can be done with hand tools, such as multi-row push seeders, full-bed tine weeders, and well-engineered wheel hoes.
- **Transplant Equipment Guild:** This is often a **scale-up** or **pro-up** guild for Market Growers depending on whether they focus more on transplanted crops or seeded crops.

EQUIPMENT FEATURE
POWER HARROW

The power harrow has important features for soil health, namely the rotation of the harrow tines *perpendicular* to the soil. Instead of inverting soil horizons, they stir the soil within the horizon. They also don't create hardpan! The roller sets depth and leaves a firm seedbed. ***Note:*** Power harrows are quite heavy and can benefit from a front bumper weight to help balance overall tractor/implement weight.

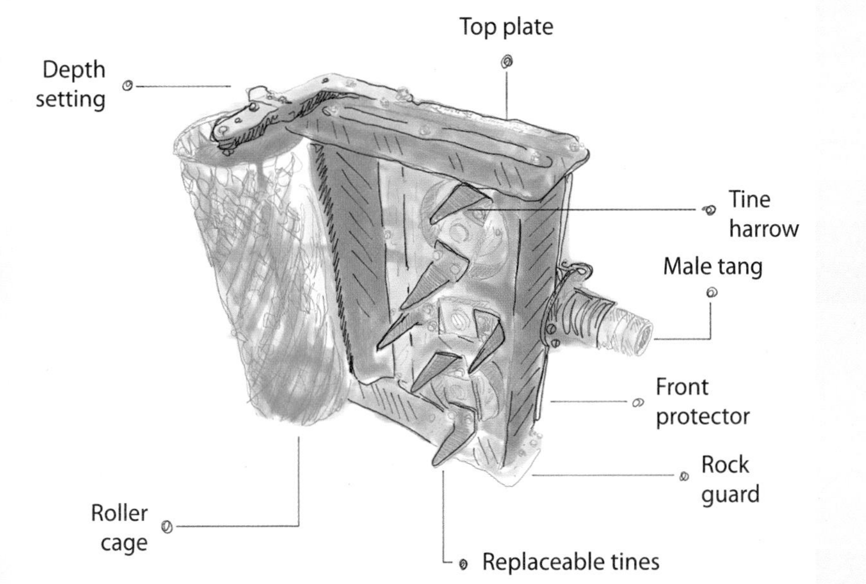

It is easier to justify a piece of equipment if it is used often. We prepare beds every week."

— Jean-Martin Fortier

DESIGN BOX

TRANSPLANT MANAGEMENT

For transplant management, growers can rely on a number of techniques, including the use of cover crops, weed barriers, poly tarps, and equipment in a series of unique systems.

ZIPPERBEDS

The Zipperbed System uses two pieces of weed barrier that meet at the crop row. This works especially well for crops that have a single row per bed and that benefit from a clean access area for regular picking, such as cucumbers, zucchini, and husk cherries.

IN SITU MULCH OR HAND-APPLIED

Hand-applied mulches include natural mulches like wood chips for paths and leaves, straw, or second-cut hay for bed tops (weed seed-free, please!). Cover crops are used for fertility, soil protection, and to outcompete weeds. They can also be grown in place as mulch. This **in situ mulch** replaces imported mulches to help crop growth by suppressing weeds, retaining moisture, and regulating soil heating. Techniques include: 1) flail mowing to chop debris and leave mulch on the bed top surface and 2) roller/crimping to create an unchopped unidirectional mulch. Take some time to make an informed decision about which type of mulching you want; doing pre-weeding is recommended for both.

ROLL-OUT MULCH

This is the popular mulching system for market gardens where a mulch-laying implement rolls out a poly or biodegradable mulch for transplanting. It is very efficient if you are mulching many rows and you can spend time on setup and adjustment. Take into consideration whether you will be removing mulch at the end of the season or using a biodegradable type. Also, certified organic commercial growers will need to consider what type of mulch is allowable for organic certification.

On the left is a Zipperbed System ready for first-year planting of melons, corn, and beans. On the right, early greens are growing between beds of overwintered rye grown as an in situ mulch.

FIGURE 18: MARKET GROWER OPERATIONS

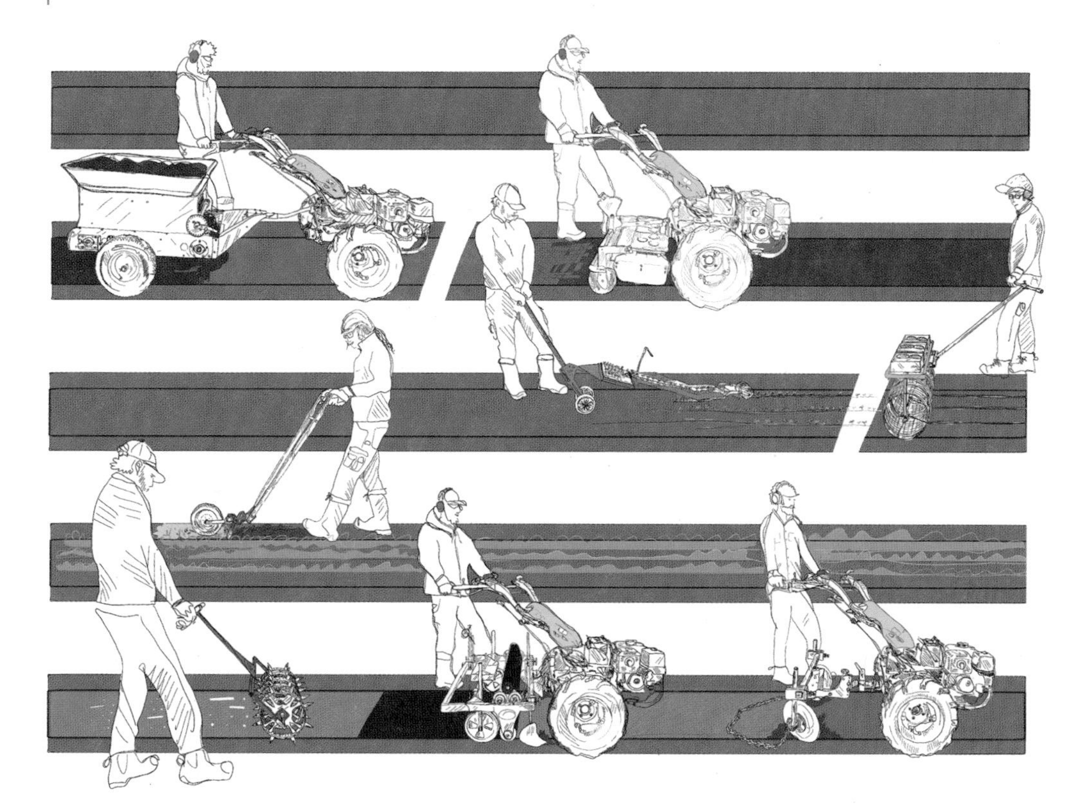

For the Market Grower, there is a string of in-season tasks to perform when moving from one crop succession to the next:

1. ***Seasonal application of compost*** *with compost spreader and/or mowing and incorporating cover crop with a flail mower and rear-tine tiller.*
2. ***Pre-weeding*** *occurs immediately following a compost application with handheld or mechanized tine weeders because compost can introduce annual weed seeds (like lamb's quarter or grasses) and perennial grass roots that can sprout. A quality and finished compost should be clean of weed seeds, but this isn't always the case or in the budget. Partially composted farm manure from a neighbor is cheap, but it's likely to be full of weed seeds.*
3. ***Final bed preparation*** *with power harrow or PDR tiller, which mixes in compost (1–3") and firms bed tops—this is considered a low-till method.*
4. ***Row management*** *using a seedbed roller (like the one offered by Johnny's Selected Seeds) firms and marks straight rows to be followed by a single-row push seeder (like the Jang or the Earthway) and/or a Paperpot Transplanter for efficient crop succession.*
5. ***Weeding of paths and bed tops*** *with hand tools rather than cultivation implements is common in market gardens. Wheel*

hoes are commonly used to maintain paths, and collinear and stirrup hoes usually manage between rows.

6. ***Reforming the bed*** *seasonally with a power ridger to improve the soil structure.*
7. ***Forming and mulching*** *bed tops with a bed shaper and mulch-layer. This reduces weed pressure, increases soil warmth for heat-loving crops like tomatoes, and holds moisture. Other mulching techniques include zipperbeds and in situ mulches.*
8. ***Dibbling*** *(making holes) and transplanting into mulch (I like using the Two Bad Cats dibbler).*
9. ***Crop protection and irrigation*** *are among the other essential systems for Market Growers. Row covers keep unwanted pests out, and irrigation is (obviously) important for crop productivity in dry seasons.*
10. ***Harvest, processing, and storage.*** *Crops are harvested and transported on a utility trailer to be washed and then put into the cold storage to await market day.*

ROW CROP FARMER (3–6 APA)

Row Crop Farmers prefer growing on flat ground divided into rowed cultivation zones measured to cultivation-equipment widths. They are *extensive* vegetable growers, planting in widely spaced crop rows, sometimes focusing on crops that are space-hungry (pumpkins), cultivation-needy (potatoes), or wholesale-friendly (tomatoes). There are really two types of row crop growers: **Row Crop Market Growers** do extensive production of many different vegetables (20–100 varieties), and **Row Crop Farmers** focus on limited (5–20) wholesale varieties (e.g., potatoes, pumpkins, garlic, and onions).

Spacing: Typical row spacings are 15" (garlic), 30" (broccoli), and 60" (melons). Even smaller crops like garlic and onions are grown with wide spacing (15"), whereas Market Growers will plant garlic with spacing of

5–6". Many Row Crop Farmers will also have crops in rows at more than 15" to facilitate weed management using mechanized equipment. However, the return per square foot is lower. The middle ground is wider spacing between rows, but the density of in-row spacing typical for market gardens (1–12" for most greens, roots, etc.) is still maintained.

Row cropping can be a single species or diversified (typical for a market garden). In both cases, more space is provided between rows to improve the speed of machine cultivation of in- and between-row weeds).

CRITIQUE OF ROW CROP FARMING

Extensive row cropping means more land is used for lower yields/acre, and the typical wide spaces between rows expose soil to erosion and compaction, facilitate losses of nitrogen and organic matter, and promote weed growth. Conversely, a wider spacing helps cultivation (providing weeding and dust-mulching benefits), improves airflow to prevent mildews, and provides easy access for harvest. Ultimately, strategic design can maximize row crop benefits and avoid its pitfalls. Some of the strategic designs are alternate land patterning, cover cropping, and companion planting (see "123 Planting Method," in Chapter 7).

TRACTOR RECOMMENDATIONS

Row Crop Farmers use multi-purpose tractors (like BCS) and row crop tractors (like Tilmor). Row crop tractors are used to pull cultivation equipment for weeding, dust mulching, and overall crop cultivation in the growing season. Used Planet Jrs are affordable (just make sure they are in good shape), but the Tilmor is also a great and budget-friendly tractor for row cropping. Thiessen Tillage Equipment is making new cultivation systems, so relying on a limited supply of used implements is no longer necessary.

An M2w-tractor is used for plowing new land, preparing fine seedbeds with a power harrow, mowing crop debris, etc. A straight-axle model with a decent engine size is fine for mostly straight fieldwork, but consider other enterprise needs as well before giving up differential and hydrostatic drive, which could be a big help with moving between plots and negotiating headlands when plowing or driving carts full of vegetables into the wash area! The hydro-mechanical clutch on popular BCS models is of benefit for stop/start in the field, which happens a lot when you have to stop often to move stones or check the quality of your seedbed preparation work.

DESIGN BOX

WEED MANAGEMENT

TYPES OF WEED CONTROL

1. **Expose (uproot) the weed:** This is done when cultivators like a basket weeder lift the young weeds out of the ground and leave them on the surface to desiccate. This is also done by uprooting weeds with a tine weeder—for white thread-stage weeds (very small weeds at the earliest stage of growth).
2. **Mechanical removal of the weed (slicing):** The weed is physically cut, removed, or pulled from the soil and dies. This is done by most cultivators (like tender plant hoes or duck-foot cultivators) on a two-wheel tractor, as well as by hand hoes and wheel hoes.
3. **Burial of the weed:** The weed is covered by soil or mulch and cannot access light, so it dies. This can be done with mini-discs, hiller discs, or spyder discs on a two-wheel tractor, or with hand tools. Discs are for bigger crops; spyder discs leave a rougher edge for better soil quality. Beet knives are another cultivator which gives some hilling; they are used with younger crops for in- and between-row weeding.
4. **Burning using flame:** The weed is withered by flame applied to the soil surface.
5. **Frying using black poly tarp:** Weeds burn and decay from excessive heat, moisture, and lack of light.
6. **Environmental conditions can control weeds:** Weeds are a product of soil conditions and improved soil horizons; aggregation and cover cropping reduce weeds.

7. **Weed-seed bank:** Shallow tillage prevents weed seeds in soil from returning to the surface during the crop cycle. Manage garden edges to reduce weed seed sources that regenerate the seed bank.

CULTIVATION ROUTINE

- Tine weeders or basket weeders *pre-weed* beds. Tine weeders *blind weed* crops (like onions), and basket weeders are *only* for small weeds among young crops.
- Tender plant hoe and finger weeders are used for the majority of crops when first planted and young, with continued use for carrots.
- Once crops get big, switch to a pathway cultivator or start hilling for crops that need it. Establishing a crop canopy smothers weeds too.

Pro Tip: *Cultivate your crops about every 7–11 days. Keeping on top of the weeds is critical.*

- Small discs can help with crops like carrots to prevent green shoulders; then use finger weeders to pull soil back to avoid leaving too much soil in the row so carrots can continue to grow unhampered by excess dirt.
- Larger crops like potatoes need more aggressive cultivation and hilling. Potatoes can be managed by switching between tine weeder passes and hilling.
- Managing crops by switching between hilling and finger weeding also works; build up the hills and kill weeds through burial, then knock the hills down and kill weeds by exposing them.

EQUIPMENT GUILDS FOR ROW CROP FARMERS

Row Crop Farmers will have a different set of equipment guilds than Market Growers because they operate *extensively.*

- **Land Prep Guild** is a **start-up** guild for opening land and row crop furrowing and hilling.
- **Single-Row Cultivation Guild** is a **start-up** guild. It includes a finger weeder/tender plant hoe combo, a pathway cultivator, and a set of spyders and/or discs that can be added to the pathway cultivator frame.
- **Fine Cultivation Guild** is a **scale-up** guild. It includes a basket weeder, tine weeder, and three-row seeder. This equipment will do pre-weeding, blind weeding, and path cultivation. Some growers opt for a better push seeder instead.
- **Soil-Building Guild** is another **scale-up** guild.

- **Multi-Row Cultivation Guild** is a **pro-up** guild. It includes multi-row tender plant hoes, multi-row finger weeders, and a spring hoe cultivator. **Add** a second Power Ox at this stage.
- **Option: Hand Tool Guilds** are for specialized jobs and are usually still **start-up** essentials, despite mechanized weeding.

"I love the old Bezzerides torsion weeders (spring hoes). The Bezzerides make a perfect combination with finger weeders for many crops."

— Jason Weston

DESIGN BOX

CULTIVATION FOR ROW-BASED PLANTING

Cultivation relies on **row-based planting** (1, 2, or 3 rows). Crop rows must be *equidistant* and *in-line* and calibrated to the equipment used to cultivate the space between rows. Row marking is essential to achieve equidistant rows. Row-marking methods include the tried-and-true stakes and string, rolling row markers (available from Johnny's Selected Seeds, Terrateck, or Two Bad Cats), and multi-row seeders (which directly seed equidistant rows). The rows for extensive farms are often 90", 60", 30", or 15" apart (typical spacings for a range of crops from pumpkins to garlic). Compare this to intensive market production, where row spacing is typically 20", 10", and 5" within the bed, 48" for a single row per bed, and 96" when one row is planted in every other bed.

To set up your cultivators (sweeps) for 3-rows at 15" spacing, first find the midpoint on your cultivator toolbar with a measuring tape and mark it. This midpoint is where your middle row of crop will pass under the toolbar. Then mark the side-rows 15" on either side. Fasten the cultivator attachments midway between all these points (which have same measurement as the actual rows on the bed top). The cultivator's width will determine how much ground is weeded. For a three-row system, usually, two cultivators are used for each row (such as tender plant hoes) to provide an overlap. These are set for a "closeness of shave" of the crop when making weeding passes. Set for a 1.5" safety zone beside crop rows. For a 12" cultivation zone, use two 7" sweeps with a 1" overlap per row.

The **cultivator track** (A) can overlap (B) for better weeding because **cultivators** (C) can be mounted using **horizontal toolbars** (D) and various **clamps** (E). The **space between cultivators** (F) can have a **cultivation buffer** (distance between the edge of the cultivator and the crop row) to ensure the vegetables *are not cut* *or buried.* Space between cultivators is usually 4", with a 1–1.5" buffer on each side, leaving 1–2" for the crop row. It is possible to increase the distance between cultivators (say 5–7") and make two cultivation passes each time, hugging one side of the crop row to get closer cultivation when crops are young, and allowing the fast and efficient between-row cultivation when they are larger. ***Note:*** Fast is better for cultivation because it does a better job killing weeds, and speed is always efficient…if, of course, you have straight rows and good soil conditions. You don't want to kill your crops in the process!

Gauge wheels (G) keep cultivators straight despite tractor maneuvering, aided by the **hitch and drop pin** (E) connected closer to the axle than with non-row crop tractors.

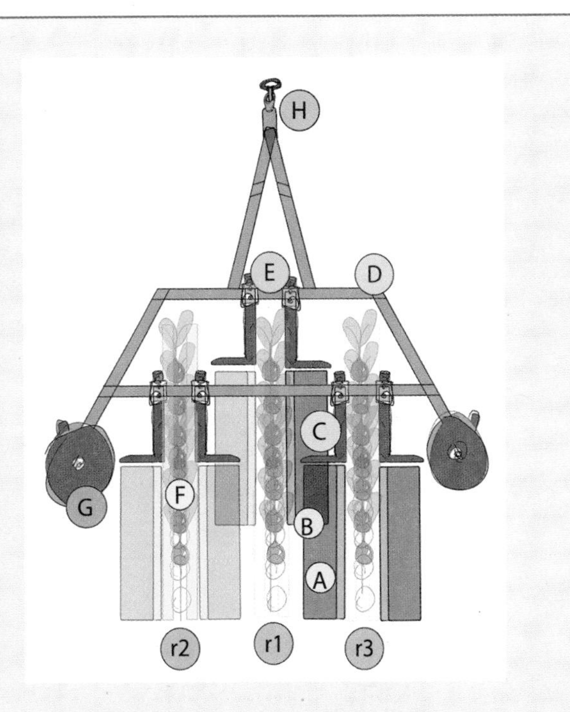

Note: Different cultivators are used for different crops, maturities, and soil conditions. For example, tougher onions can handle finger weeders that cultivate both in- *and* between-row. But you'll need tender plant hoes to cut close to fragile crops like spinach.

DESIGN BOX
ROW CROP TRACTORS ARE DESIGNED FOR CULTIVATION

Row crop tractors have their hitch very near to the axle, and this prevents implement wandering when making tractor direction adjustments as you cultivate. For instance, with an M2w-tractor, if you were to drive too close to the left, and you want to steer right, the equipment behind you would actually jog left before it followed your tractor to the right. For 4-wheel tractors, this is best managed with mid-mount cultivators (like the Allis-Chalmers G), but for pull-type implements, you need the hitch to be as close to the axle as possible to minimize this steering swing to less than an inch, rather than several.

FIGURE 19: SUMMER CULTIVATION CYCLE

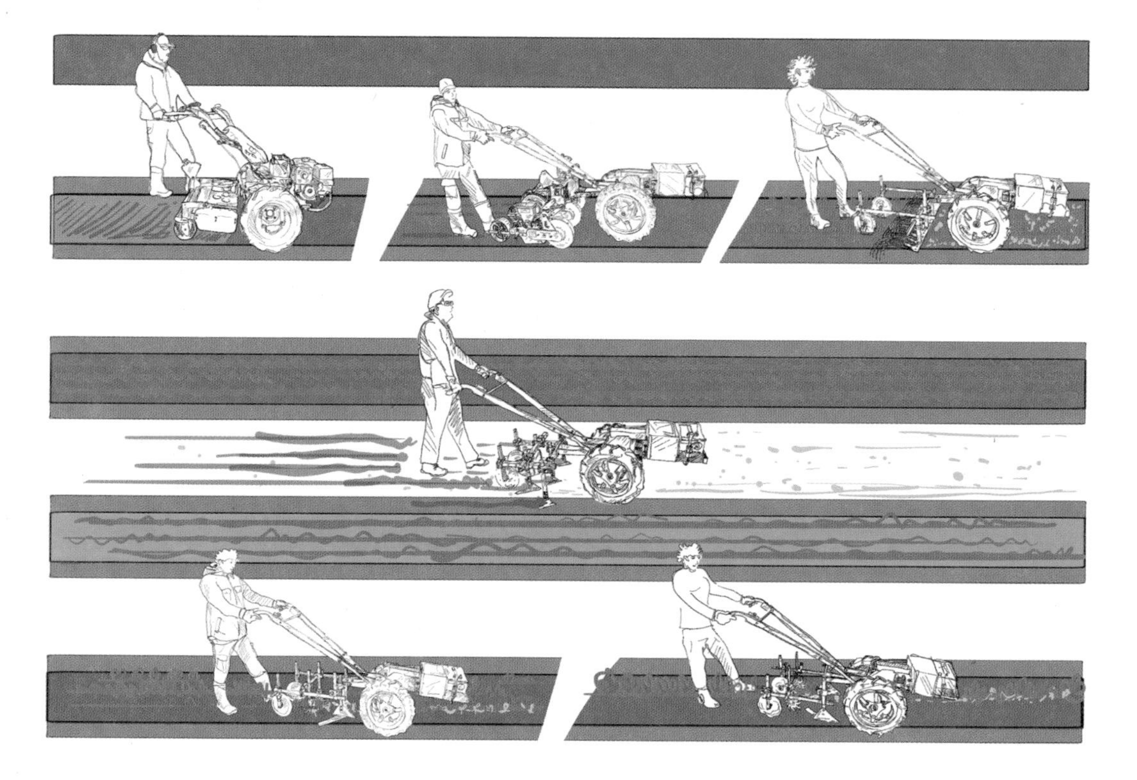

1. ***Field preparation*** *using a rear-tine tiller or power harrow.*
2. ***Pre-weeding*** *using a tine weeder before crops are seeded.*
3. ***Field seeding*** *of crops using 3-row seeder places seed at precise row spacing.*
4. ***Blind weeding*** *using tine weeder to cultivate white thread-stage weeds.*
5. ***Pathway cultivation*** *keeps alleys clean of weeds using S-tine or chevron cultivators.*
6. ***Row cultivation*** *with finger weeders manages in-row weeding, and tender plant hoes remove between-row weeds; this combo works well together. A beet knife can be used to throw soil into the plant row to bury weeds as well as for between-row weeding in a single pass for some cropping situations.*
7. ***Hilling cultivation*** *with hillers (for crops like potatoes that need more mounding) or spyders is done for more or less aggressive in-row and between-row weeding for beans.*
8. ***Bed edge cultivation*** *is done along mulches and raised beds using cultivators to pull the weedy soil away from the edge and then put it back.*

BACK-TO-THE-LANDER (3–15 APA)

The Back-to-the-Lander's goal is multiple resilience-building enterprises that suit their property, which tends to be larger, more remote, and often consists of more varied land types, such as woodlots, fields, gardens, ponds, etc. Back-to-the-Landers tend to mix *extensive* and *intensive* enterprises readily, including smaller kitchen gardens and orchards alongside larger storage field crops, hay fields for livestock, and woodlots for forest products.

TRACTOR RECOMMENDATIONS

A Back-to-the-Lander may benefit from larger, more versatile tractors with many more functions than are needed by Backyard Gardeners, Market Growers, or Row Crop Farmers. Hillside properties will need heavier models, and larger acreage will benefit from the maneuverability of differential drive or even hydrostatic drive. A good option is something like the BCS 770 or 779 (or similar). If you are more remote, however, a standard transmission and clutching system might be more practical because they are easier DIY repairs. Additionally, throwing in a row crop tractor for your large garden wouldn't go amiss.

EQUIPMENT GUILDS FOR BACK-TO-THE-LANDERS

Many equipment guilds can be useful for Back-to-the-Landers, where self-sufficiency is the name of the game, and equipment must suit the variety of landscapes these tractors will need to operate within. There are many equipment options for this tractor user.

- **The Forester Guild** is a **start-up** guild for those with *a wooded property.* Also, for those who need to cut back the woods to clear land for gardens or hay or construction. For those clearing land, the log splitter could be switched for the brush mower.
- **Land Prep Guild** is a **start-up** guild for those who want *a big garden.*
- **Homestead Guild** is a **scale-up** guild for those with more *lawn and water management* needs.

- **Soil-Building Guild** is a **scale-up** guild for those with *larger gardens* and helpful for *hay fields* and *lawns* too.
- **Grassland Guild** is a **scale-up** guild for those who are *tending animals* and *making hay*. Great for those who want to be self-sufficient for meat and/or want to produce their own "weed-free mulch" for gardens.
- **Land-Clearing Guild** is a **start-up** guild for those needing to *clear land*. It includes a brush/flail mower, stump grinder, and chainsaw.
- **Row Seeding/Weeding Guild** is a **pro-up** guild for those wanting to grow a serious *storage crop garden* to fill their cellar.
- **Most Hand Tool Guilds** are useful at **start-up** and **scale-up** phases.

A GLIMPSE AT FOUR SEASONS OF DOING-IT-YOURSELF

Back-to-the-Landers are known for their self-sufficient and DIY lifestyle. With the right complement of equipment, those folks can get most jobs done with just one engine. Especially since jobs come along in different seasons, it is more practical to add useful implements rather than a second tractor.

1. Move material to and from the field with the **buddy cart**, or use a **utility trailer** for large loads that might require a flat bottom.
2. Keep your equipment and vehicle clean and spray out your cider press, maple syrup pan, and animal bins with a **pressure washer.**
3. Keep your fence lines clean or cut your hay with a **sickle bar mower.**
4. Keep your woody debris cycled with a **wood chipper**, turning it into mulch and path material.
5. A **snow blower** can help with many small trails or even do a small driveway.
6. A **log splitter** is a great way to affordably heat your house, cabin, or other outbuildings. It feels great to fuel your sugar shack with firewood from your own bush lot.

FIGURE 20: BACK-TO-THE-LANDER OPERATIONS

There is always a need for chopped wood at the sugar shack in March. Moving seamlessly from splitting logs, chipping branches for mulch, mowing sapling in trails, and hauling sap is a great example of an equipment complement for small maple sugar bush operators.

GROUNDSKEEPER AND LANDSCAPER (VARIOUS APA)

The Landscaper and Groundskeeper use most of the same equipment as these two are fundamentally providing a *service for* property owners instead of property owners *DIYing it*. This includes installing landscape features like gardens and orchards, or maintaining these features, as well as keeping all those ubiquitous lawns mowed.

TRACTOR AND EQUIPMENT RECOMMENDATIONS

These users need a larger and very multi-functional tractor that can handle all sizes of equipment. A hydrostatic transmission would be helpful for all mowing and other applications. Tractors like the BCS 770 or 779 (or equivalent) would be good choices. Most equipment guilds are useful to landscapers as they scale-up to be able to do more types of custom work.

FIGURE 21: GROUNDSKEEPER OPERATIONS

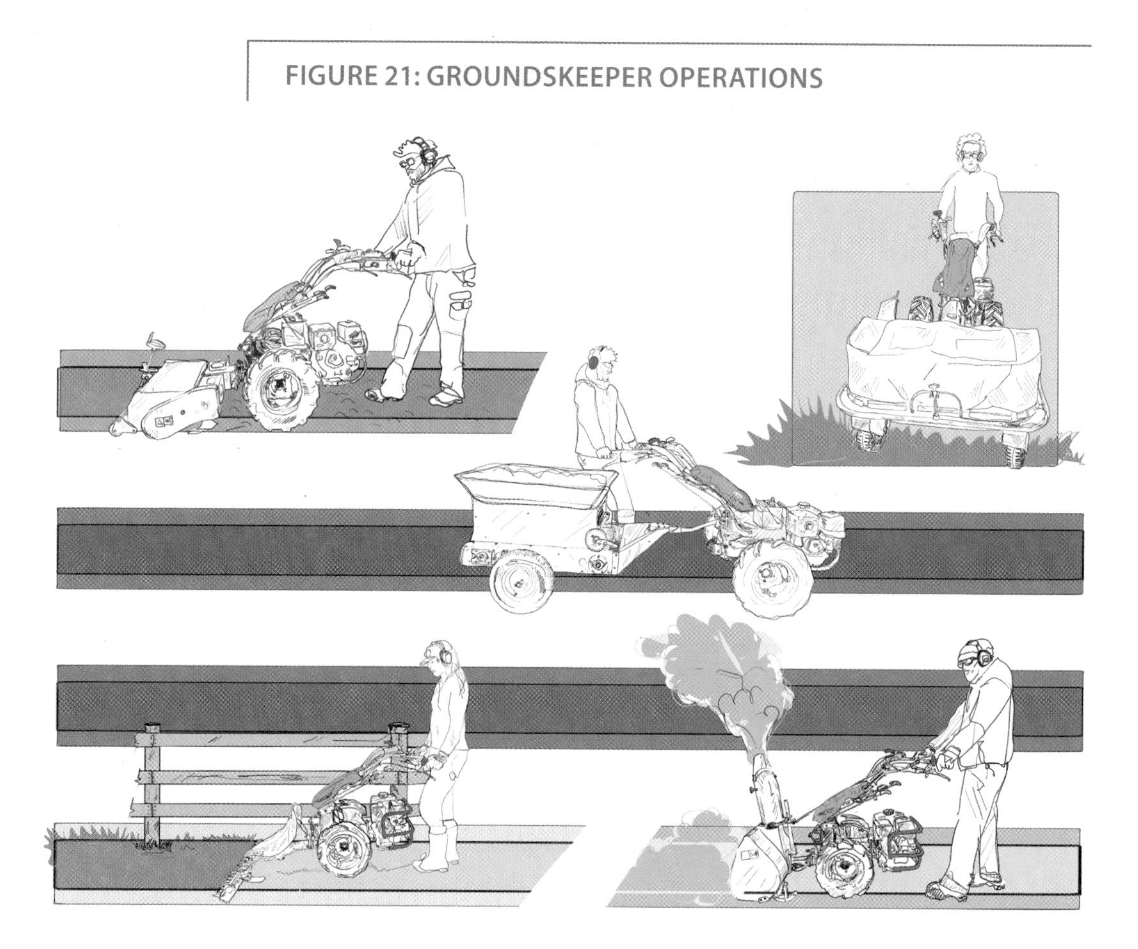

The Groundskeeper/Landscaper performs a variety of tasks—from dethatching, compost spreading, and lawn mowing to sickle bar mowing along fence lines and winter groundskeeping with a larger two-stage snow thrower.

EQUIPMENT FEATURE

LANDSCAPER ESSENTIALS

- **The swivel rotary plow** is more versatile for landscapers who might work on *sloped* or flat ground and for maneuverability in *tight spaces*, such as property lines, fence rows, foundation walls, and other situations where returning into the same furrow is much easier.
- **The power ridger** is a great implement for landscapers because it allows you to make a number of unique features in the landscape. In addition to its typical bed-forming utility, it does unique jobs, like trenching for buried irrigation lines, flower bulb planting, and creating a depression for base material under new patios and walkways.
- **The dual-action sickle bar mower** is a great implement for groundskeeping because it can mow under fences and tree lines with its T-shaped mowing system.

The rotary plow is very useful for making new landscape features. It is also quite maneuverable and can even be used to create more sinuous lines, such as this circular feature for a food-forest maze. Here, the plow is being used as an edger to sharpen the grass boundary.

GRASSLAND FARMER (8–13 APA)

The Grassland Farmer is focused on the production of hay and the maintenance of pasture for livestock. Grassland Farmers often manage a number of hay and pasture fields with rotational grazing of small-scale farming of cattle, sheep, and other livestock. There are other applications for the tractors and equipment of the Grassland Farmer; see the lists for orcharding and restoration, below. ***Note:*** Grassland Farmers can benefit from a small four-wheel tractor for moving hay bales.

TRACTOR AND EQUIPMENT RECOMMENDATION

A hydrostatic drive is recommended to facilitate turning and speed adjustments when operating mowing equipment. Grassland Farmers may also benefit from heavier models with increased traction. This might include wheel options that provide better grip on slopes, such as studded wheels. ***Note:*** When mowing or on slopes, *the operator should always be downslope,* with the handlebar adjusted to an ergonomic height.

A compost spreader is a practical addition for improving pasture yield. Otherwise, the implements for a Grassland Farmer are very specific to hay production. The Grassland Equipment Guild is a **start-up** guild for anyone who is specializing in grassland production on a decent scale. But it would be considered a **scale-up** guild if grassland production was an *added* enterprise.

After passing with the sickle bar mower, this hay rake gathers cut and dried grass and legumes and windrows them for the baler to pick up.

TREE NURSERY (3–6 APA)

The Tree Nursery can make great use of two-wheel tractor applications for *intensive* seedling and *extensive* sapling management (for future resale as potted or bare-root trees). Management of nurseries can be done with two-wheel tractors up to 3 to 6 acres. ***Note:*** Tree nurseries can benefit from a small 4w-tractor for harvest and supply management.

TRACTOR AND EQUIPMENT RECOMMENDATIONS

Most M2w-tractors suited to market growing will work in the nursery. Additionally, a row crop tractor can be useful for the cultivation of young tree seedling beds. The techniques and equipment suggestions that follow are for a unique system that maximizes two-wheel power for tree management. This considers three types of tree nursery work: direct tree seeding into beds, transplanting small seedlings (3–6") and growing them out to medium bare-root trees (12" to 24"), and "holding over" of larger potted or bare-root stock (as a safe way to keep stock healthy before landscape jobs or sales in early spring or over the summer and winter for future sales).

The rear-tine tiller can re-prepare beds adjacent to tree nursery rows that were planted in trenches made by the power ridger and filled with the rotary plow.

EQUIPMENT GUILDS FOR TREE NURSERIES

Some of the equipment and operation cycles will be familiar to Market Growers and Row Crop Farmers. The focus for a nursery will be on what is different in the operation cycles as it pertains to growing perennial trees for sale. The possibilities this opens for highly profitable small-scale nurseries in suburban and small rural properties are huge! Tree nurseries are one of the most profitable forms of small farming.

- **Tree Planting Guild** is a **start-up** guild for tree nurseries. They can use the power ridger to form planting trenches, the rotary plow to fill them back in around tree whips (4 ft trees), and the power harrow to help manage the space between tree rows for seeding cover crops.
- **Soil-Building Guild** is a **scale-up** guild that allows growers to more easily manage compost additions to improve the soil tilth on bed tops for smaller tree seedlings and use the flail mower to manage inter-row cover crops.
- **Row Seeding/Weeding Guild** is a **pro-up** guild that allows tree nurseries to maintain tree stock grown from seed or transplanted as small (2–6") starts by allowing thorough between and in-row cultivation. This equipment guild uses row crop tractors.

FIGURE 22: TREE NURSERY OPERATIONS

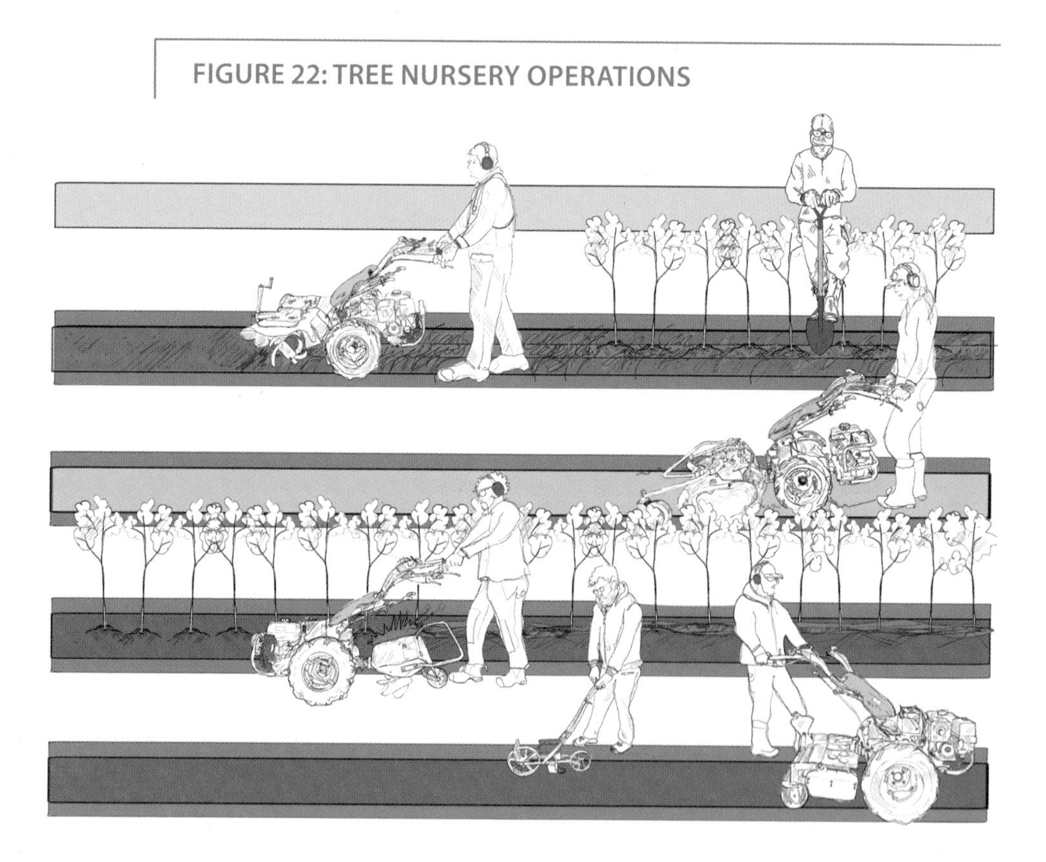

FARM FEATURE

ZACH'S NURSERY SYSTEMS

INTENSIVE MULTI-ROW TREE GROWING

1. **Build and re-prepare seedling beds** using the popular power ridger, tiller, and power harrow operation cycle.
2. **Seed nursery beds** using a 3-row seeder or push seeder with a row marker. Straight rows are as critical for trees as for vegetables—even more so since trees grow over several years.
3. **Transplant small bare-root trees** into single or double rows. Three rows are often too much for trees; a little more room is helpful for mechanized weeding. One row is very efficient but can use too much land.
4. **Consider cover cropping** by hand or using a broadcast seeder to cover crop all the unused ground after a few weeding cycles to create an in situ mulch over the nursery area. **Note:** *Don't forget to install rodent guards on your trees!*
5. **Weed seedling beds** between tree rows with a row crop tractor and then seed or plant into green manure.
6. **Lift trees** using hand tools, a two-wheel tractor with a root digger or subsoiler, or a typical four-wheel tractor tree-harvester setup.

EXTENSIVE SINGLE-ROW TREE GROWING

1. **Create large-stock nursery beds** using a power ridger to make trenches down the center of a bed. Place potted stock or large bare-root whips in a tight single row (only 3" to 12" between stems).
2. **Maintain large-stock nursery beds** using a rotary plow to not only fill the trench back in but also for subsequent passes on either side in weed sweeps to keep in-row weeds and shoulder weeds from getting established.
3. **Lift large tree stock from a nursery bed** using a subsoiler set to the far right side of the tractor; cut along each side of the tree to help loosen it. ***Note:*** Large-stock nursery beds are only used to hold tree stock over one or two summer/winter cycles. Root establishment will still be minimal, allowing easy lifting *without* the heavy four-wheel equipment used for lifting in most nursery operations.

DESIGN BOX

INTEGRATING NURSERY AND MARKET GARDEN

Tree nurseries can easily be partnered with market gardens by intercropping young trees with cover crops and vegetables. Here, it is done with the *Compost-a-Path Method.* Permabeds are built to have 30" bed tops, and paths are also 30" wide for easy access and to allow cover cropping. Cover crops are mowed to create green alleys for access when digging fruit trees for sale and distribution. The cover crops can be turned over when needed for new crops. Trees are grown to a certain size before harvest and either replanted with more spacing for larger stock or sold as smaller whips (2–4' single-stem trees). Larger trees are grown for another summer and held over the winter in trenches filled with loose soil to sell in early spring as bare-root trees that can be tagged for customers and then harvested before leaf-out for distribution.

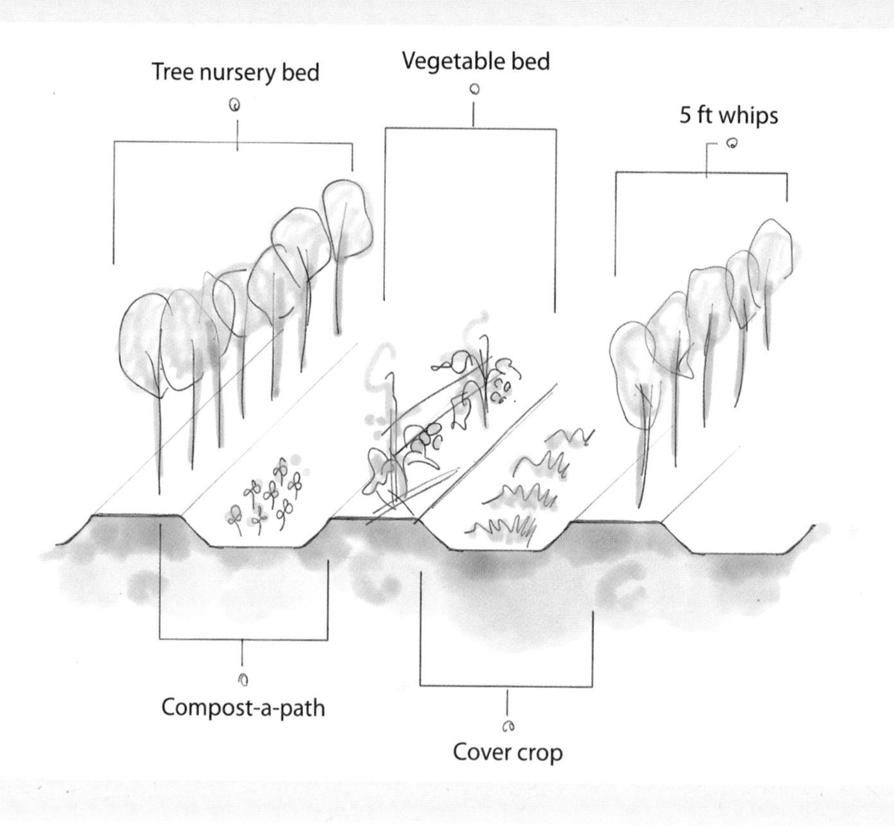

EDIBLE ECOSYSTEM MANAGEMENT

TREE PLANTING FOR ORCHARDS AND REFORESTATION

The tree planting equipment and other essential equipment for tree nurseries are also helpful for many different perennial enterprises and operation cycles for establishment and maintenance: orchards, reforestation, ecosystem restoration, silvopasture, Permaculture, etc. Here, we use **Edible Ecosystem Design**. Covered in detail in my book *The Edible Ecosystem Solution,* and discussed in the next chapter, it is an approach that can be applied to all enterprises thus far discussed, and the **Permabed System** is a model for ecosystem design that uses organized plant patterning.

The application of ecosystem design to any perennial project is mainly that the trees are planted farther apart and mixed with other desirable perennials. *The intended goal is the harvest of fruit, timber, nuts, and berries and the fostering of ecosystem services.*

EDIBLE ECOSYSTEMS

Edible Ecosystem Design is the process of creating land management systems that mimic wild ecosystem form and function but are organized and efficient for management. A "food forest"— a term being used more and more often these days—is a good example of an edible ecosystem. But really, an edible ecosystem can be anything from a wild tallgrass prairie to an integrated orchard market garden. This latter example, also called a **Permaculture Market Garden**, is covered in detail in my book *The Permaculture Market Garden*, which includes designs for diversified plantings of fruit, berries, and herbs as edible hedges between fields in garden, hay, and row crops. We will explore this further in the next section that showcases the Permabed System in detail for two-wheel tractor use.

FARM FEATURE

TERRA VIVA

High up the side of a mountain in the Dolomites in northern Italy, Terra Viva, an organic, Permaculture-inspired market garden is exploring a multi-faceted approach to growing food for their local community. Here, the steadfast Market Grower techniques of cover cropping, crop rotation, pre-weeding, and tarp culture are found alongside diversified production with annuals, perennials, and animals. The farm employs animals in rotation with crops, grows edible hedges of herbs, and has a new chestnut orchard *on keyline,* a method that helps to slow, spread, store, and harness water in the soil for productive cropping on sloped land using the land's topography. In addition to all this, they intercrop trees as a source of carbon and for its soil-building capacity. Chipped limbs and branches are layered in the greenhouse, in paths, and on fields and allowed to decompose as in situ mulches; they are balanced with nitrogen-rich inputs, such as manure and green manure crops, like vetch.

ABOVE PHOTO CREDIT: TERRA VIVE

FIGURE 23: EDIBLE ECOSYSTEM OPERATIONS

Managing diversity is often seen as difficult, but organized design and small-scale equipment can provide efficient management of multiple enterprises within the same plot. For instance, micro-hay works great between orchard or afforestation rows. Permabeds can be pattern-planted in fruit, berries, and herbs and integrated as edible hedges between hay, crops, and market vegetables. When managing diversity, small equipment is appropriate, such as the sickle bar mower for managing under fruit trees.

1. *Organized edible hedges with hay or other agriculture plots between.*
2. *Sickle bar mowing underneath orchard trees in edible hedges.*
3. *Use a power ridger to make furrows for tree planting.*
4. ***Plant trees into the trench** and use a shovel to deepen the trench for taproot trees, like pecans.*
5. ***Maintain the shoulders of the trees against aggressive weeds** using a narrow rear-tine tiller (18") just to keep back aggressive weeds early on.*
6. ***Mow the space between trees** using a flail mower. The lawn mower is also great because it collects clippings that can be used as mulch around trees.*
7. ***Fill the newly planted trench and hill the trees** for weed management using the rotary plow, building up the bed around them.*
8. ***Spread compost and flail mow cover crops** in the space between trees with the compost spreader and flail mower (not pictured).*
9. ***Cultivate row crops** between orchard trees with row crop tractor and cultivation equipment.*
10. ***Chip branches** from young fruit trees into mulch with the wood chipper/shredder (not pictured).*
11. ***Harvest fruit** into buddy carts or utility trailers (not pictured).*

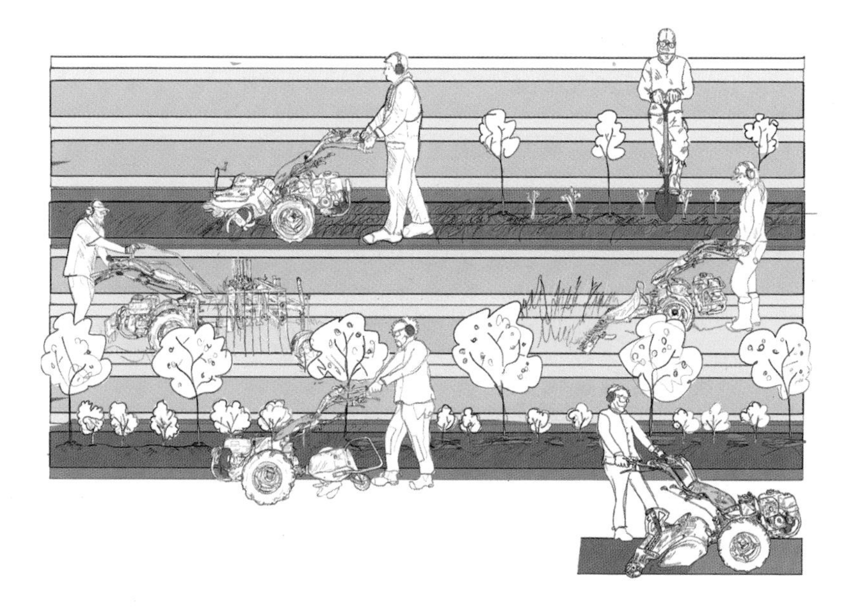

DESIGN BOX

ECOSYSTEM DESIGN FOR TERRACES

The woodlands in Tuscany lend themselves to truffle production because of the abundant beneficial relationships that exist there among mycorrhizal fungi and specific tree species: pines, oaks, and hazels. These relationships help the forest thrive and provide food for people. **Using Ecosystem Design and restoration principles,** these wild tree species can be grown as hedges on earthen terraces while still maintaining other valuable crop production *and* providing truffle agri-tourism! Planting in earthen and stone terraces (like those used across the Mediterranean for thousands of years) is making a comeback for small-scale farming today. These terraces help to avoid erosion where food is being grown on steep slopes, and these gardens are easily managed with two-wheel tractors because they fit in tight spaces.

The steep slopes around Genoa, Italy have been contour-terraced for small-scale agriculture to produce the world-famous Genovese basil for pesto. Just down the coast, the intensive basil farms in the working town of Pra are where the basil is *actually* grown. Here, two-wheel tractors are a natural fit on small hillside plots and intensive greenhouses. The design on the adjacent page pays homage to this landscape while using the Permabed System for patterning of annuals and perennials.

Traditional olive growing needs space for netting, ladders, and carts; this is planned for on the terraces, as well as in flatter fields. Planting designs using AMP (Alternate Maturity Patterning) put adjacent garden beds in early maturing vegetables. Once harvested, they leave space open for efficient olive harvest.

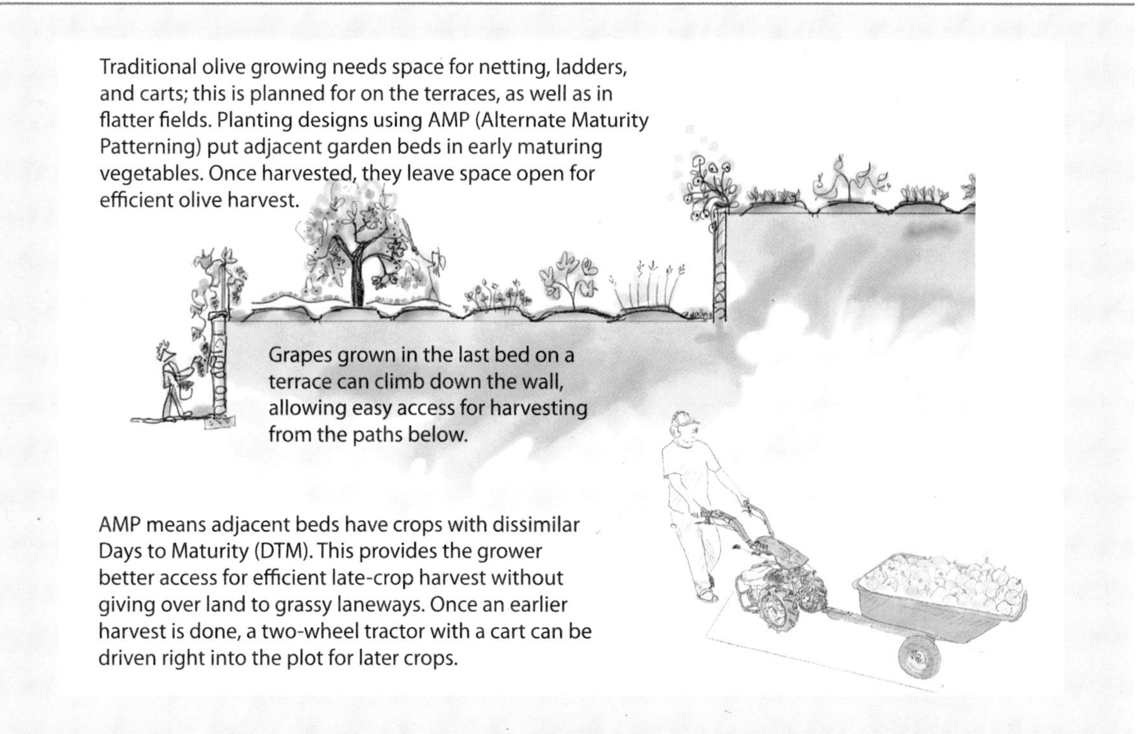

Grapes grown in the last bed on a terrace can climb down the wall, allowing easy access for harvesting from the paths below.

AMP means adjacent beds have crops with dissimilar Days to Maturity (DTM). This provides the grower better access for efficient late-crop harvest without giving over land to grassy laneways. Once an earlier harvest is done, a two-wheel tractor with a cart can be driven right into the plot for later crops.

Roadside hedges offer valuable materials: bamboo is transformed into grape tomato trellises, and saplings that are chipped into mulch. Chip-mulched paths on terraced properties reduce erosion and reflect sunlight to reduce rapid daytime slope heating.

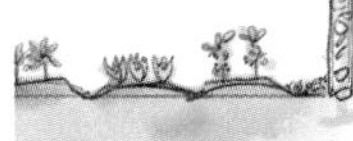

Retention walls planted with companionship plants help the entire property. Although they don't facilitate harvest, they provided needed pollinator habitat and pest-deterrents.

Power harrows or rear-time tillers are used on Permabeds for bed preparation in the greenhouse. Flail mow and tilled debris mitigate disease. Diversified crop rotation also reduces disease pressure while offering further harvests for regional sale and export—with the same equipment used in the basil production and pesto processing.

FARM FEATURE

LA SCOSCESA

This market garden in the Toscano hills is growing specialty crops like saffron alongside a plethora of popular market garden veggies. Lorenzo grows all this in soil full of rocks while also rebuilding ancient terraces. The maneuverability and the low impact of the two-wheel tractor are perfect for negotiating narrow and steep farm lanes, yet work effectively to build up the soil and prepare it for seeding and planting.

Intensive greens growing in terraces at La Scoscesa. Lorenzo's saffron, which is harvested from the crocus flower's stigma, is incredibly beautiful growing in the beds—and tasty too!

Chapter 6

Permabed System For Two-Wheel Tractors

Growers get one shot per year for seasonal success, and achieving it is more complex when managing biodiversity on the farm. The **Permabed System** is a land management design for intensive and extensive growers of annuals, perennials—and even animals—that uses *ecosystem design,* in other words: mimicry of natural systems for high productivity and resilience in farms and landscapes. This system employs a complement of *landscape layout, guild design, crop rotation,* and *input/output* management techniques blended as a complete multi-year management cycle.

The entire system presented (or subsets) can be adopted and adapted for different small-scale growers. Our focus here is an in-depth case study to tie together this book's material and show *how equipment can be applied to a complete production system.*

Note: A complete review of the Permabed System and Ecosystem Design is beyond the scope of this book. For more information, see my earlier books, *The Edible Ecosystem Solution* and *The Permaculture Market Garden.*

PERMABED SYSTEM 101

A Permabed is a raised garden bed that is formed and reformed, but never destroyed. Its unique architecture improves soil life functions, and its fixed place in space allows the patterning of diverse food plants into edible ecosystems designed for gardens, farms, and landscapes.

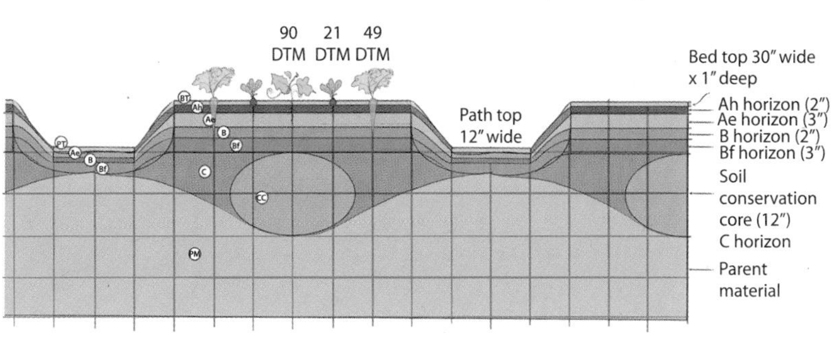

The Permabed is designed to conserve soil life and integrate edible biodiversity in a layout organized for efficiency and profitability.

PERMABED SYSTEM DESIGN PRINCIPLES

The Permabed System is designed for efficient, organized, and profitable edible, and useful biodiversity. Diversity is difficult to manage; historically, monocultures and limited polyculture have been the standard for food production. The Permabed System uses **ecosystem design** to organize landscapes and relies on crop guilds and planned management cycles.

The following **design principles** outline the Permabed System succinctly:

- **Edible Ecosystem Design** integrates annuals, perennials, and even animals within landscapes in an organized and efficient manner to improve profit resilience for your enterprises.
- **Permabeds** are permanent raised beds with unique designs for an ecosystem approach to growing.
- **Bed Permanence** means Permabeds are made and reformed, but never destroyed.
- **Garden Environment Mapping (GEM)** relies on bed permanence to map the unique hydrology, soil, and weed pressures of distinct beds for improved plot-scale and bed-scale management.
- **Soil Conservation Core** is the soil layer found in Permabeds underneath the production bed top. In this zone, the soil is not disturbed by deep tillage and has well-formed soil aggregates, which serve as habitat for beneficial microorganisms.
- **Organizational Land Patterning** organizes farms and other landscapes, laying out lanes, alleys, and plots to create an overall design that integrates annuals, perennials, and animals within the same plots. Organized units include beds, triads, and plots.
- **Crop Guilds** are designed to improve companionship among different plants and the soil ecosystem.
- The **123 Planting Method** and organizational units like **triads** (3 beds) are used to assemble annual or perennial crop guilds.
- **Alternate Maturity Patterning (AMP)** is a guild design and land patterning technique in which crops of differing maturity days are always adjacent.
- **Umbrella Management** is a design technique that ensures crops chosen for guilds can be managed easily together. Examples: All need to be transplanted, or all have heavy fertility needs.
- **Guild Crop Rotations** organize crop guilds within a schedule of when they are grown in a particular bed and plot.
- **Management Cycles** are multi-year plans for tasks, equipment, and crop rotation, such as in situ compost generation using the Compost-a-Path Method.
- **Compost-a-Path Method** maximizes *compost generation* in paths, improving bed turnover speed for new crops. This in situ compost is then applied back onto the bed tops as fertility every three years.
- **Task Schedules** show which tasks are done when, specifying the equipment used to complete operation cycles.

DESIGN BOX
PERMABED DESIGN PRINCIPLES

1. **Annual (or perennial) guilds** include *three or more crops* within the same bed and/or triad.
2. **Triad design** relies on a single **key crop** coupled with appropriate companions; the key crop is either in the middle or in both outer beds. The key crop usually has the longest Days to Maturity (DTM) within the guild and requires the most management. **Example:** Squash has the greatest space needs in the late season, so you design around this characteristic.
3. Crops can be grown as a monoculture, in parallel (different crops are in their own plots, but adjacent) or patterned (different crops occur proximate within beds and in adjacent beds as guilds). **Ecosystem Design** has the greatest diversity of these methods, with annuals and perennials using the Permabed System to organize this diversity for efficiency and profit resilience.

Many planting points of diverse edible plants are managed in organized rows with in- and between- row patterns in a Permabed.

Ecosystem Design mimics wild woodland layers where ground covers, herbs, bushes and trees are organized to enhance net photosynthesis, which can equal higher yield.

Zipperbed method uses weed barrier to slowly allow perennials to fill newly weed-free soil.

A single Permabed is transitioned to an edible perennial ecosystem along a driveway or as an edible hedge within a market garden plot.

Alternate Maturity Patterning makes sure Permabeds are intensively managed while allowing extensive crop needs, like squash sprawling.

Annual vegetables are organized across a triad (an organized unit consisting of three Permabeds) to help partition resources and space.

Early DTM

Late DTM

Early DTM

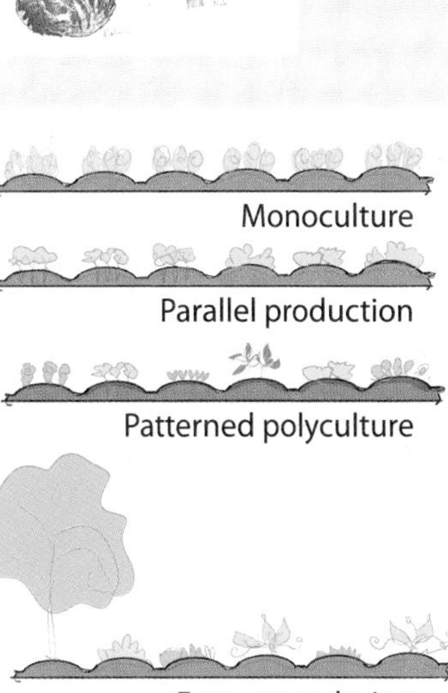

PERMABED ARCHITECTURE

For two-wheel tractor operation within this system, let's start at the beginning, with a single bed.

FLAT-GROUND, RAISED BED, OR PERMABED

The practical question of flat-ground or raised beds is a useful one for growers to ask. As a reminder: flat-ground growing is popular because it is easier to mechanize, especially for cultivation. Market Growers, in particular, like raised beds because they have looser soil for transplanting, more root depth available for plants, better drainage, improved solar warming, and less compaction of the soil profile. A Permabed is a raised bed; however, because

DESIGN BOX

PERMABED SOIL DYANMICS

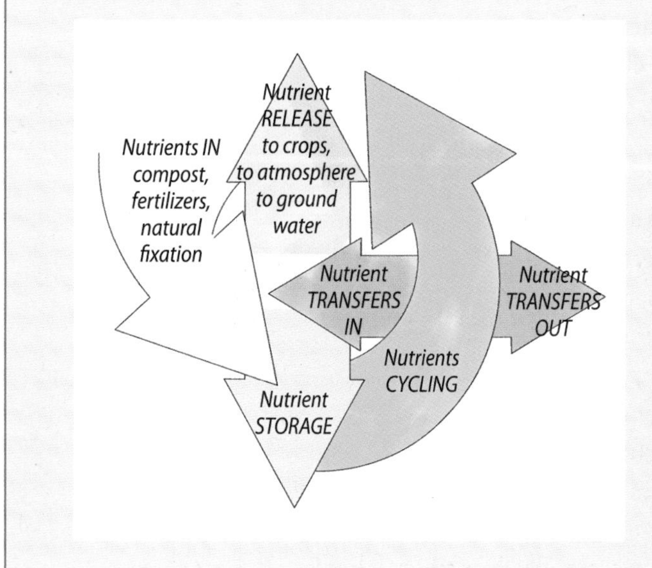

Permabeds are designed to capture, fix, store, cycle, and release nutrients, water, and organic matter with minimal loss. This is because they build organic matter, improve pathways for resource movement, and increase storage capacity with well-aggregated soil. This results in high levels of beneficial soil life, such as nitrogen-fixing bacteria and mycorrhizal fungi. By building Permabeds, we are setting up a garden framework to help efficiently reduce tillage impact, prioritize compost and cover crops, maximize crop diversity, and maintain a soil conservation core!

it is formed and never destroyed, it has many other benefits. These include *enhanced* long-term soil life health, connectivity within the soil and between crops and soil organisms, and improved nutrient and water cycles. They also allow much more complex guild crop design and rotation.

QUALITY OF RAISED BEDS

There are many types of raised beds, with varying quality and benefits. Not all raised beds are made the same, and when we understand our goals and operation practices, we can determine equipment use that matches our intentions.

SEASONAL RAISED BEDS

These are made by tilling between tractor wheels and fluffing, aerating, and *temporarily* improving the planting soil. The bed shape (and height) is enhanced by tractor wheels and humans packing down the path. These actions do help growers seasonally, but within a month, the soil will settle back down. Another version entails forming with a bed shaper or rotary plow for a 2–3" raised soil height. But with successions of tillage, this will *also settle to flat* within a growing season. Additionally, the whole field is sometimes plowed for disease/weed management, and beds don't have a permanent location. On the other hand, the benefits of seasonal raised beds include being easy to build and cultivate and allowing easy tractor clearance.

PERMABEDS

Permabeds have a maintained, raised form because they are usually formed higher and designed for a stable soil structure and well-defined soil horizons. These beds are also never plowed under, and growers can incorporate routine soil-building processes and dedicated equipment use that strengthen soil life's ability to stabilize their own environment for mutual benefit. Reforming beds is done using power ridgers or rotary plows, alongside other equipment. Mechanization of operations requires more forethought. Growers can still manage for disease/weeds in Permabeds (typically done by plowing the field) by using specialized reforming processes that don't harm the soil conservation core. Crops connect quickly with soil organisms to outgrow pests, diseases, and weeds and develop natural pest defenses, such as the rich lipids that form in their leaves.

A) *Flat-land growing is easy to mechanize for draft implements.*

B) *Seasonally formed raised beds can provide seasonal benefits, but they quickly settle down to the same level as their paths.*

C) *Permabeds have a distinct raised architecture that has integrity over time and holds its form.*

A LOOK AT PERMABED ARCHITECTURE

Permabeds have architecture with distinct zones: bed top, shoulders, path, and core. Their measurements should fit your equipment, and wheel tracks are adjusted accordingly. A Permabed's total width is center-of-path to center-of-path (see image). The bed top is measured from one top edge of bed shoulder to the other. Architecture can include unique arrangements to accommodate equipment on the bed top, straddling the bed with wheel extensions, and even in the path using a Compost-a-Path Method for efficient management of field debris and cover crops. ***Note:*** Bed lengths will vary according to enterprise needs (see Figure 24).

FIGURE 24: PERMABED ARCHITECTURE

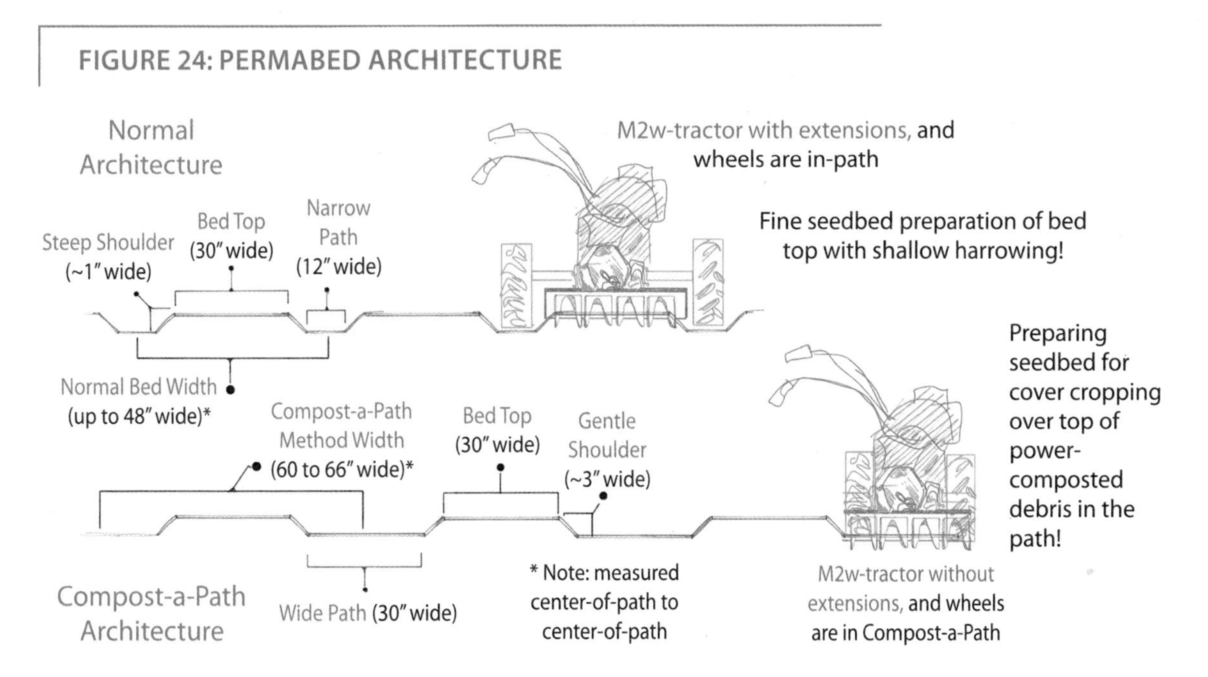

***Note:** The shoulder is not part of the bed top width because, as beds get higher, a distinct angle of repose is necessary; different shoulder widths will suit different management methods.*

DESIGN BOX

HOW EQUIPMENT MAKES BED ARCHITECTURE

- The rotary plow or tiller and hiller/furrower can easily form a 12" path and raise the bed 3 to 5 inches.
- For a wider bed system, the power ridger is perfect. The land needs to be plowed and tilled first. For a 5-foot (60") bed (center-of-path to center-of-path), mark out all the lines of the path centers.
- Then, one pass with the power ridger leaves a 35" path space between the newly formed beds, which are now roughly 26" on the bed top and raised 5–8".
- After a pass with a PDR tiller or power harrow on the bed top, the path will narrow to 30"; the bed top is widened to 30", and the height reduced by 2".
- For a 66" bed, with wider paths and wider/more gradual shoulders, the power ridger is run down marked lines 66" apart, leaving the 34" trench and forming a bed top that is ~32"

wide; this can either be softened to 30" with gentle shoulders for a taller bed with 30" bed top, or multiple passes can leave a wider and lower bed. To widen the path, multiple runs with a power ridger with the baffle set to the highest setting (allowing more soil to jettison out) are needed. A garden rake or other bed-top tool can help to set the desired width more precisely (especially for initial forming after primary land preparation of plowing and tilling a new field).

- ***Note:*** There is more discussion about this equipment and technique in later chapters.

WHERE TO PUT YOUR WHEELS?

Permabeds can be designed to suit all equipment scales. Hand tools work well with Permabeds of many different heights. If tractors are used, you will need to find a way to minimize compaction, conserve soil conservation, and make them work within your operational logistics plan. Mostly, this means using S4-tillage and other soil conservation/building methods, but it also is a question of where the tractor puts its wheels. This is put into practice via the wheel setup.

Where will the tractor wheels be when making passes in the garden?

- **In-path:** Use wheel extensions and make the bed height lower than the clearance under the tractor when wheels are in the path.
- **On-shoulder:** Use wheel extensions and a wider shoulder design to get more clearance with wheels on the shoulder.
- **On-bed:** Make wheel tracks narrow so it is completely erased by tillage passes because wheels are immediately in front of the tiller or power harrow.

Will all tractor operations use the same wheel location strategy or different ones?

- **In-path:** Used for *tillage* operations because the bed is susceptible to compaction when tilled, especially if more than one pass is needed for bed preparation.
- **On-shoulder:** Used for *cultivation* passes because wheel tracks can be erased but compaction to growing crops is limited.

- **On-bed:** Used for *mowing* operation because bed top is more stable after 6–8 weeks in a crop or cover crop. Also used for tillage operations like power harrowing because wheel tracks are erased.

Will different tractors be used?

- **Multi-functional tractors** have lower clearance, and will require lower bed height if **in-path** is chosen. Generally, M2w-tractors are used on the bed top, so considerations of bed height and extensions are necessary if wheels are in the paths.
- **Row crop tractors have** higher clearance for **in-path** wheel placement, but they provide much better cultivation control when on **bed top**. Generally, row crop tractors are used for flat-ground crop cultivation, so this would be a modified use and would benefit from wider bed tops.
- ***Note:*** Ultimately, all of these approaches will work—which you choose will depend on your own preferences and designs.

COMPACTION RESULTING FROM DIFFERENT TASKS

Part of the reason for configuring a tractor to have wheel extensions is to prevent the compaction from the tractor on the bed top itself. Different tasks have different impacts. Let's consider these below.

- **Plowing and Bed Forming** have *high short-term impacts*, but *very low long-term impacts*, and building Permabeds will allow lower compaction for years to come.
- **Final Bed Preparation** using a tiller or power harrow is *medium impact* because they should be operated at a shallow setting, according to S4-tillage principles.
- **Seeding and Cultivation** are *low/medium impact* because row crop tractors are light. Especially in later years, when soil structure is stable, seeding and cultivation will be *low impact.*
- **Post-Harvest Tasks**, like flail mowing, are *medium/low impact* because the bed is stabilized by crop roots and debris. Leaving debris and under-sowing to cover crop, then mowing when the ground is partially frozen and allowing the cover crop to emerge is *very low impact.*

WHEEL EXTENSIONS AND TRACTOR CLEARANCE

When using wheel extensions to span a Permabed, you must consider your tractor's clearance so the bed top isn't rubbed by the tractor's underside. The clearance of a multi-purpose tractor is about 5.5" to 8". This 8" is under the lowest point under the base plate of wheel brakes when using 5" × 12" wheels (22" tire diameter). The tractor underbelly or front bumper may scrape bed tops when Permabeds are over 3–5" high. The clearance underneath a compost spreader is also a limiting factor for the highest Permabeds, so keep bed height in mind when planning for spreader use. Row crop tractor clearance is 9" for the Planet Jr BP-1 and 13.5" for the Tilmor Power Ox, but no easy over-the-counter extensions are available.

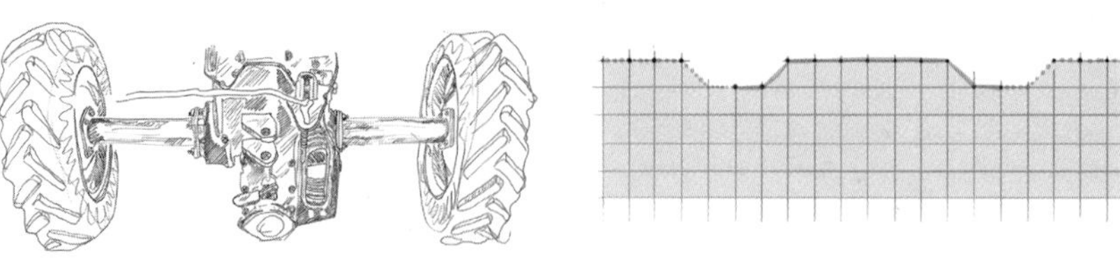

Adjusting your tractor wheel track so wheels are in-path is advisable for reducing compaction, but doing so sets a limit on Permabed height and width.

ADDING WHEEL EXTENSIONS

There are three different extension widths commonly available: 4.5", 5.5", 16", and 22" (total additional track width). Available in pairs, they require an extension accessory for each wheel that is mounted between the axle and the wheel. Also, by turning the wheel hubs around, an additional 3 or 4 inches can be added when using any of the wheel extension sizes.

Example: The BCS 739 (5" × 10" wheels) has a 34" track width when using 22" extensions (enough to clear a 30" bed top with wheels on-shoulder,

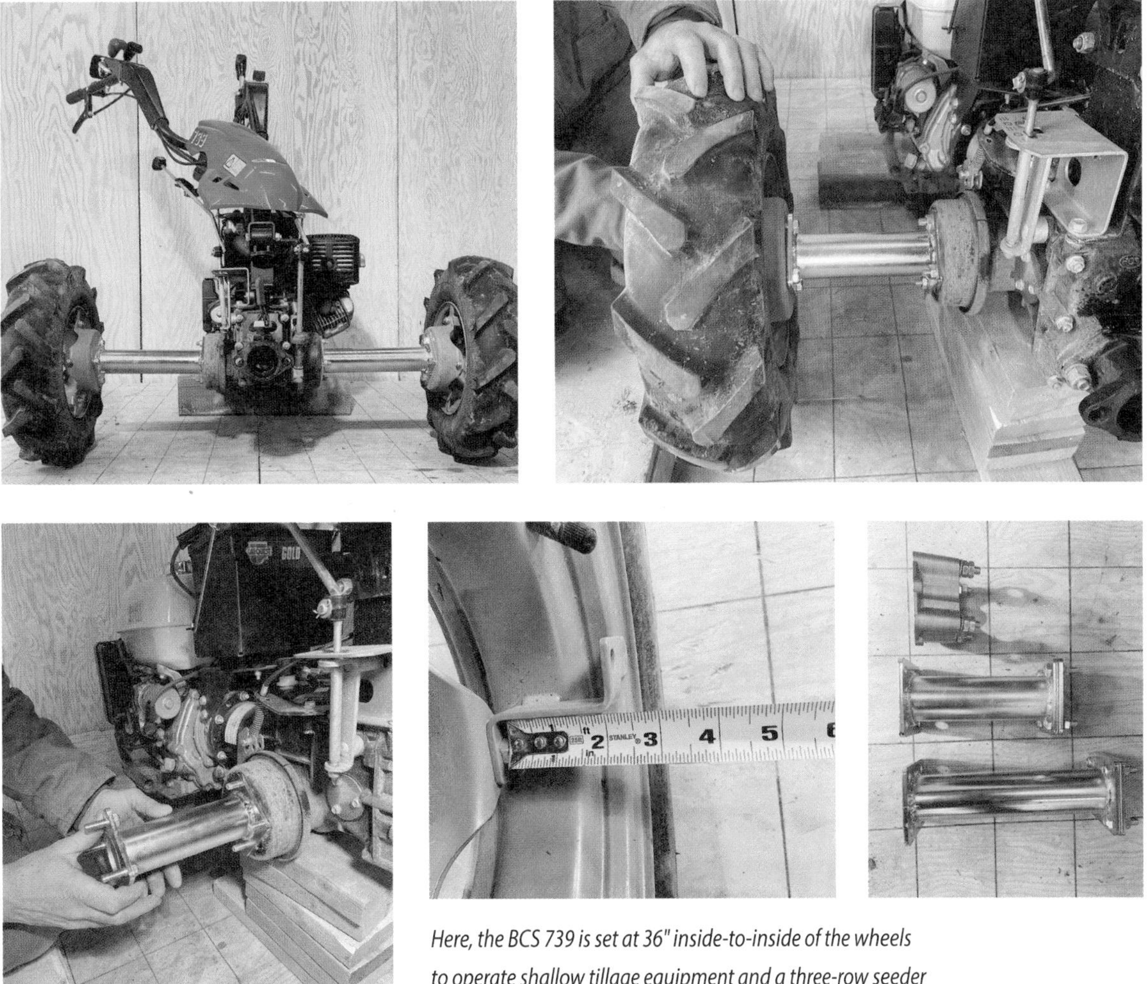

Here, the BCS 739 is set at 36" inside-to-inside of the wheels to operate shallow tillage equipment and a three-row seeder without compacting the Permabed.

and 37" when turning the wheel hub. This puts the wheels in-path, even with a 3–5" Permabed.) ***Note:*** The Compost-a-Path Method offers mechanized efficiency, even for Permabeds that are 5–9" high because wider paths allow more cover crops to be managed on the flat ground while beds grow crops; then crop debris from beds is also applied to the paths for flail mowing and power composting with earthworking equipment. The wider paths mean two-wheel tractors can easily manage the path with the same equipment used to manage the bed top. More on this, below.

WARNING: When using extensions at the widest settings, DON'T use barbells and wheel weights. There may be a risk of tearing the transmission gear if you get in a situation where the tractor has poor traction. This results in too much torque. These wide settings are ONLY used for Permabed (low slope) operations of fine seedbed preparation, cultivation, and mowing.

Note: For mowing (lawn, flail, brush), the wheels should not extend past the width of the mower itself. For sickle bar mowing on steep slopes, longer extensions can be used. Extensions can also be used for plowing.

PERMAPLOT DESIGN

Growers always need organized field layouts for the seasonal tasks of weeding, watering, and harvesting, but diversified production requires *more* organization. Permaplots are the largest scale of diversified organization and contain within them many other units of space organization for crops trees and gardens.

ORGANIZATIONAL UNITS

We have already talked about many organizational units: acre is one, beds and rows are others. The Permabed System has **10 field organization units**. These organizational units help with the layout of new plots, organizing equipment tasks, and planning crop rotations. Different enterprises will rely on various organizational units, and defining yours will help you to better understand your equipment-scale. Whether you primarily work on a bed-scale, triad-scale, or plot-scale, you will need to adjust your tractor and equipment choices.

FIGURE 25: LAND ORGANIZATION UNITS FOR DIFFERENT ENTERPRISES

Organized Land Units (width in Feet)

Block Plot Triad Bed

108 ft 36 ft 12 ft 4 ft

Measurements here are based on 48" Permabed (30" bed top, 12" path, and 6" shoulders

Bed Length for specific Enteprises

Normal Bed Widths	
48" to 66"	Row Crop Farmer (up to 200 or 300 ft beds)
48" to 66"	Market Grower (25 to 200 ft beds)
48" to 66"	Back-to-Lander (25 to 100 ft beds)
36" to 48"	Homesteader 15 to 75 ft beds
36" to 48"	Garden (5 to 15 ft)

		Permabed	15 feet	25 feet	50 feet	100 feet
TRIAD ONE	Rotation #1	Tomatoes	Left Bed	TRIAD	PLOT	BLOCK
		Early Transplant	Center bed			
		Peppers	Right bed			
TRIAD TWO	Rotation #3	Garlic (2nd Year)				
		Summer M. Veg				
		Garlic (2nd Year)				
TRIAD THREE	Rotation #5	Potatoes				
		Peas				
		Bush Beans				
TRIAD FOUR	Rotation #7	Brassica TP				
		Early TP Mix Veg				
		Brassica TP				
TRIAD FIVE	Rotation #9	Early Mix Veg				
		Early Carrots				
		Early Mix Veg				
TRIAD SIX	Rotation #2	EMB -> Garlic (1st yr)				
		Storage Carrots				
		EMB -> Garlic (1st yr)				
TRIAD SEVEN	Rotation #4	Sweet Potatoes				
		Early TP Mix Veg				
		Husk Cherries				
TRIAD EIGHT	Rotation #6	EMV-> OwCc				
		Early Carrots				
		EMV-> OwCc				
TRIAD NINE	Rotation #8	Squash & Zuchinni				
		Corn & Pole Beans				
		Cucumbers & Melons				

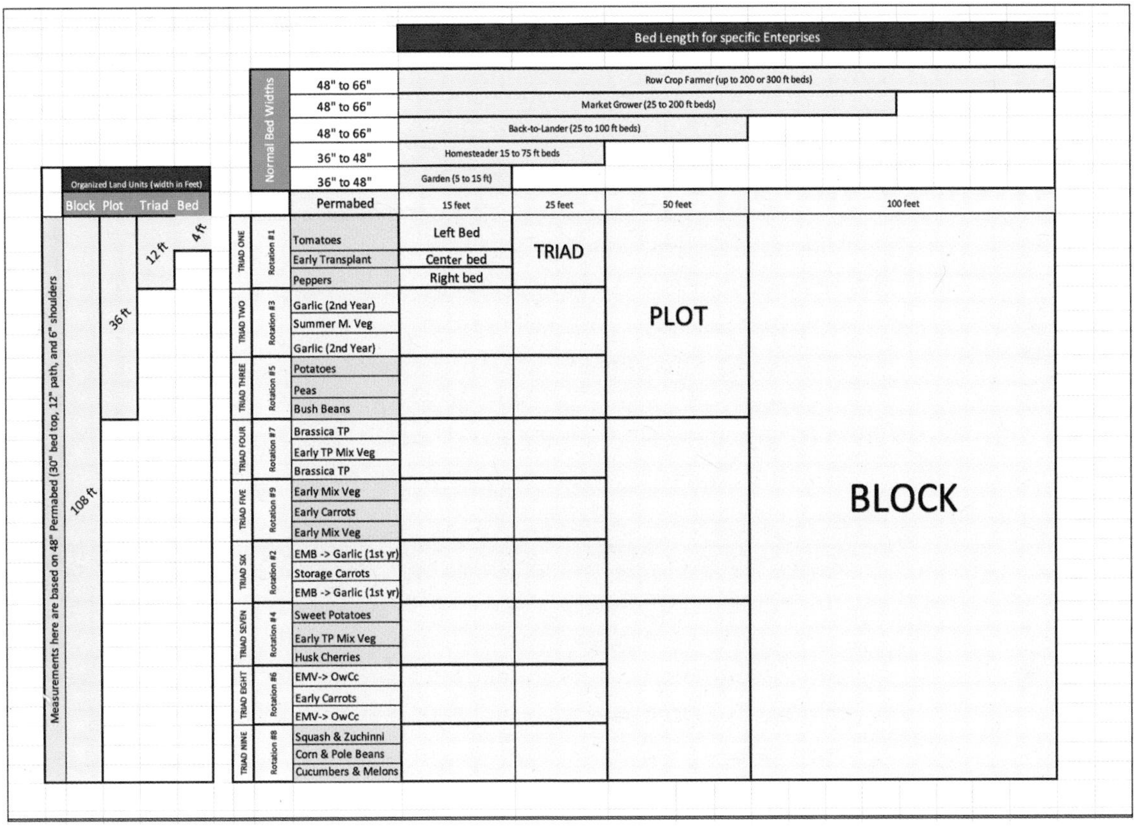

This concept design shows the best organizational units for different enterprises.

Note: *This table is included here primarily to reference bed length. Crop rotation content can be best understood by reading* The Permaculture Market Garden *and taking courses at EcosystemU.com. Email EcosystemU@gmail.com for a larger image file of this diagram*

UNITS: SMALLEST TO LARGEST:

- **Planting point:** Exactly where a seed or plant goes in-row.
- **Row:** Runs the full length of the bed, always equidistant between, with 1, 2, 3, or 5 (or more) rows.
- **Spot:** A 5-foot width of bed that a distinct annual or perennial guild can occupy, e.g., a plum tree with companions. Allows patterning of guild spots along the full bed.
- **Permabed:** One single bed measured from center-of-path to center-of-path. With the Compost-a-Path Method, a single Permabed is best

understood as one Permabed and the entire right-hand Compost-a-Path. Can be dedicated to annuals or perennials. Remember, Permabeds can be any length, but they are usually 15', 25', 50', 100', 200', or 300'.

- **Sub-bed:** A longer bed usually divided into equal parts. Usually made to facilitate crop rotation when beds are too long for a certain crop, like cilantro i.e. 25 feet of a 100 foot bed.
- **Triad:** Three Permabeds grouped together as a unit for guild design. Can be dedicated to annuals or perennials. Has the same length options as Permabed. Is the basic unit of guild crop rotation.
- **Sub-triad:** Same as sub-bed.
- **Plot:** Usually three triads (total of nine Permabeds). Annuals or perennials are dominant.
- **Blocks:** Usually three plots for larger field organization.
- **Permaplot:** Always includes annual and perennial triads together in the same plot.

PERMAPLOT LAYOUT

A Permaplot has annual and perennial crops integrated and patterned by beds and triads—usually at a ratio of 2 or 3 annual triads for 1 perennial triad. **ANA** is the short form for an annual triad and **PERA** is short for a perennial triad. So, the basic Permaplot pattern is this: PERA/ANA/ANA/ANA. This is 1 traid in perennial and 3 triads in annuals. Then it repeats. Perennial triads can also be planted to be *emergent* (with trunked trees) or *regenerative* (canes and bushes that regenerate easily from underground growth or coppiced species).

FIGURE 26: PERMAPLOT ORGANIZATION

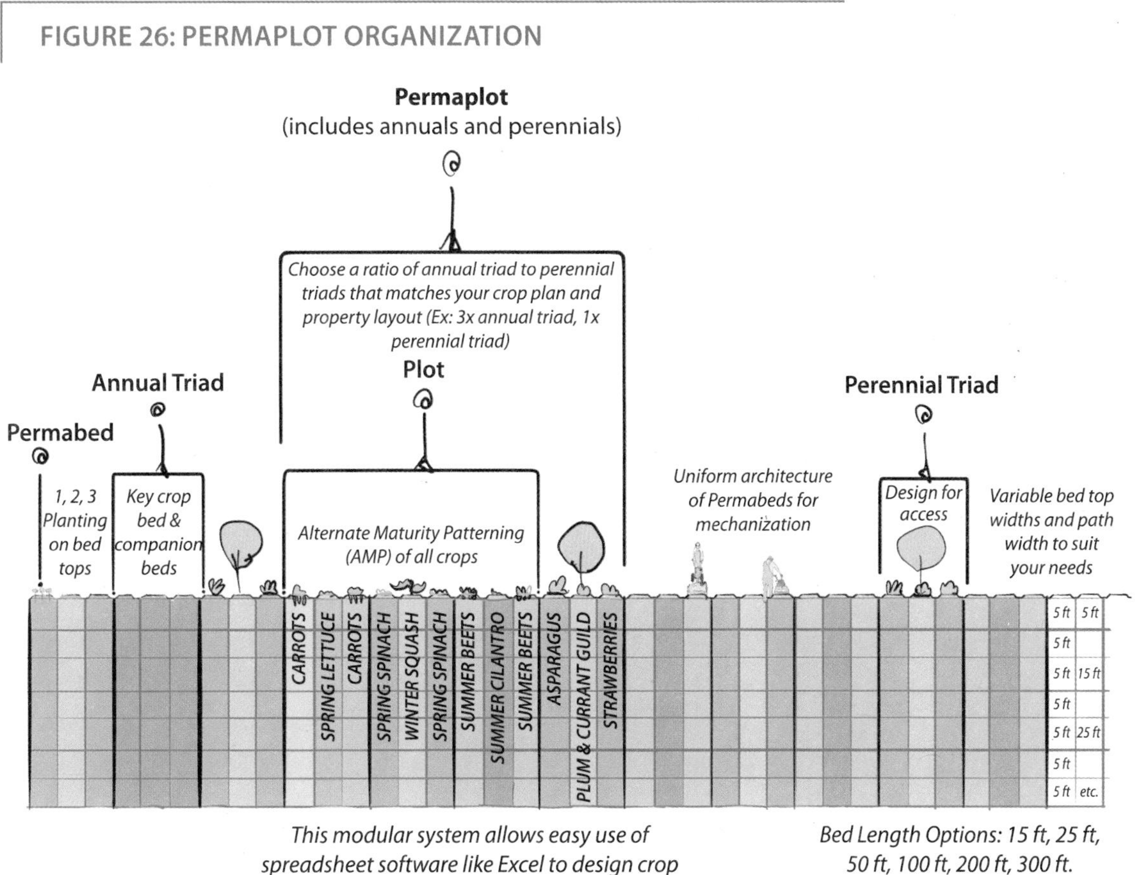

The Permaplot is the basic landscape-scale ratio of annual-to-perennial triads (ANA-to-PERA ratio).

DESIGN BOX

ZACH'S ESSENTIALS (3+ ACRE BIODIVERSE FARM OR EDIBLE LANDSCAPING)

These six implements are great for growers with over 3 acres—especially for those managing a biodiverse farm or managing landscape businesses that have many varied needs for equipment to manage soil, crops, cover crops, and fertility. A diverse farm will also benefit from carts, trailers, wood chippers, irrigation, etc., but the implements can also be pulled and powered

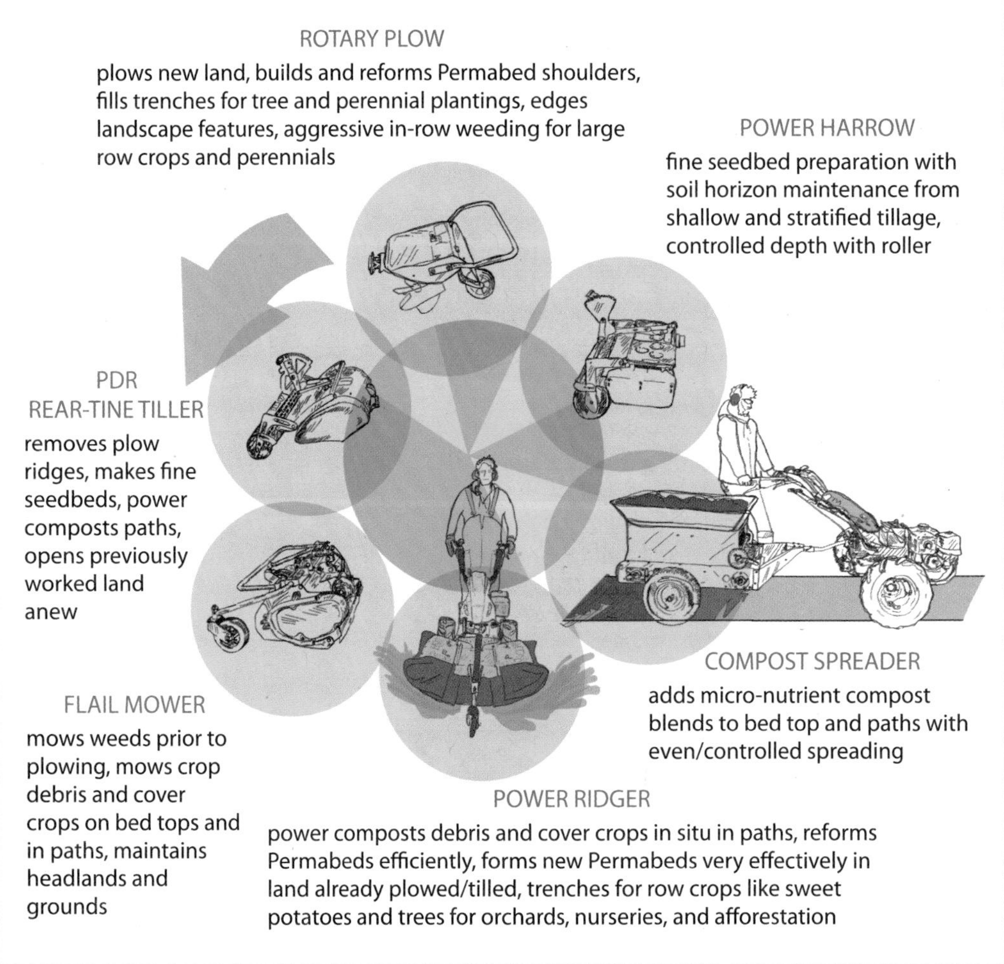

ROTARY PLOW

plows new land, builds and reforms Permabed shoulders, fills trenches for tree and perennial plantings, edges landscape features, aggressive in-row weeding for large row crops and perennials

POWER HARROW

fine seedbed preparation with soil horizon maintenance from shallow and stratified tillage, controlled depth with roller

PDR REAR-TINE TILLER

removes plow ridges, makes fine seedbeds, power composts paths, opens previously worked land anew

COMPOST SPREADER

adds micro-nutrient compost blends to bed top and paths with even/controlled spreading

FLAIL MOWER

mows weeds prior to plowing, mows crop debris and cover crops on bed tops and in paths, maintains headlands and grounds

POWER RIDGER

power composts debris and cover crops in situ in paths, reforms Permabeds efficiently, forms new Permabeds very effectively in land already plowed/tilled, trenches for row crops like sweet potatoes and trees for orchards, nurseries, and afforestation

The professional grower should ***start-up*** *with a rotary plow, PDR tiller, and flail mower. Later, the logical additions are a power ridger, compost spreader, and power harrow.*

by other equipment like ATVs, trucks, compact tractors, electric pumps (and sometimes contract work, depending on your goals and scale). But when it comes to essential earthworks, mowing, and fertility, growers should be well-equipped to maximize their two-wheel tractor.

Tractor User	**Tractor Cost** (USD)	**Equipment Guild Cost** (USD) *(3 implements)*	**Safety Gear, Maintenance Tools & Supplies Cost** (USD)	**Totals** (USD)	**Annual Income and /or Savings** (USD)	**Full-Time Equivalency** (how many people employed)
Backyard Gardener (.5 to 1 acre in garden)	**BCS 710** ~ $2,000 or **BCS 722** ~ $3,300 (smaller engine, less features, great for just gardening) *~$1,500 used in good condition*	**Basic Garden Guild** ~$2,825 *(less $1,725 if flail mower will be purchased later as market grower scale-up), ~$1,800 used*	~$200 to $600 depending on how much essential safety gear and garage tools you already own	$3,500 to $6,725	$15,000 to $25,000 in food savings and small sales	~.5 to 1.5 full-time equivalents
Market Grower (1 to 3 acres)	**BCS 732** ~$3,500 (entry level for professionals) *~$2,300 used in good condition)* *or* **BCS 739** ~$4,500 (great professional tractor) *~$3,100 used in good condition*	**Land Prep Guild** ~$6,360 *(less $995 if tiller was purchased when Backyard Gardener)* **Soil Builder Guild** ~$6,100, *~$4,000 used for each*	~$200 to $600	$6,600 to $11,100	$45,000 to $300,000 in gross sales + food for farmers	~1.5 people to 6 full-time equivalents
Row Crop Farmer (3 to 6 acres)	**Tilmor** ~$2,700 ~$1,000 for *a fixed-up used Planet JR*	**Row Seeding/ Weeding Guild** ~$3,100, *~$2,100 used*	~$400 to $900	$13,500 to $23,300	$150,000 to $500,000 in sales + food for farmers	~3 to 6 full-time equivalents

Note: Pricing is in USD and is current as of 5/2022.

The Row Crop Farmer Equipment total includes the Market Grow Land Prep Guild and Soil Builder Guild + the Row Seeding/Weeding Guild.

From left to right in the image: Tarp culture pre-weeding of beds for early crops the next spring; garlic beds were quickly planted in October, with a fast turnover process: crop debris removal, power harrow, and dibbling. Crop debris in paths will be flail mowed next spring as an in situ mulch over the cover crop that will be over-seeded into the debris in spring.

COMPOST-A-PATH METHOD

Compost-a-Paths are a *Permabed variation* with these **immediate benefits**: improved plot access, mechanization of path management, efficient cover cropping, enhanced bed reforming, refined crop debris logistics, more crop successions per year, and, as the name implies, in situ compost generation *in your paths*. **Long-term benefits:** improved soil structure and soil food webs, water and nutrient cycling, and a resilient approach to pest management.

This system *is not for all* growers but *will be revolutionary for some* who adopt its designs. It is an *extensive* management system (requiring more space), but this space is used very effectively and efficiently, making it *intensive!* In this system, wider paths are created, and this is where most bed-top debris is composted, cover crops are grown, and two-wheel tractors travel to manage the path as they would a bed top. As concerns budgeting space/time/energy, the Compost-a-Path Method is a net positive for serious growers who design for its use.

COMPOST-A-PATH PRINCIPLES

- **Think about** your path as you would a bed top.
- **Compost-a-Path widths** (22" to 36") need to accommodate a tractor between beds for task mechanization.
- **Equipment options:** All bed top equipment for Permabed Systems can be used on both paths and bed tops.
- ***Note:*** Different path and top widths require *either* two-wheel tractor setups OR adjustment of wheel width.
- Path receives crop debris from **spent bed**,[1] leaving bed top clean for timely succession planting.
- **Debris** can be mowed as-is and left on the bed top when desired, but debris is NEVER removed to compost piles.
- **The disease management** viewpoint is that moving debris around fields and the poor pile and windrow systems typically seen on farms are not

1 A **spent bed** is a Permabed that has had its crop harvested and thus has debris that needs to be cleaned up so another succession can be planted.

sufficient for disease management. Better to create dynamic compost systems in situ.

- **The crop debris** in paths is often mowed to serve as mulch over top of newly over-seeded cover crops, and weed barrier in paths is used for composting both crop debris and later cover crops by covering it over and allowing moisture and heat to build up. You can continue to walk on the paths as the wide 30" format and the effective mechanized management (such as flail mowing) create a low profile for composting debris and cover crop material.
- **Nitrogen-fixing cover crops** balance carbon-rich debris and create in situ compost in paths.
- **Crop rotation** in adjacent beds follows a three-year management cycle based on heavy, medium, and light fertility-feeder debris-creator crops. A fertility-feeder crop is one that requires more energy from the soil feeding off the soil's reserves; a debris-creator is one that provides a lot of material to the soil, which, in the long term, becomes a new source of fertility. These crops are usually one and the same; most crops fall on a spectrum of being either low, medium, or high feeders and providers. By creating a crop rotation that balances these, we can make sure more nutrients are available naturally from debris-creators by the time fertility-feeders are planted again. As such, a 3-year management cycle is designed so that at year three, all crops will have fully decomposed, even in drier climates, and, with Compost-a-Path Method, the decomposed material is in the paths, so it can then be moved onto the bed tops to feed future crops.
- **Cover crops** in paths help improve composting and change path ecosystems—attracting predatory insects and fostering soil life, which limits disease.
- **Grow-and-mow** weed management in paths allows annual weeds to be mowed as opposed to being tilled or cultivated. This is a layered approach to growing; we add onto the land layer after layer, rather than doing too much tillage.

***Pro Tip:** Equipment use is interchangeable when the bed and paths have the same width. **Example:** Mowing both bed-top rye and/or tomato debris in paths.*

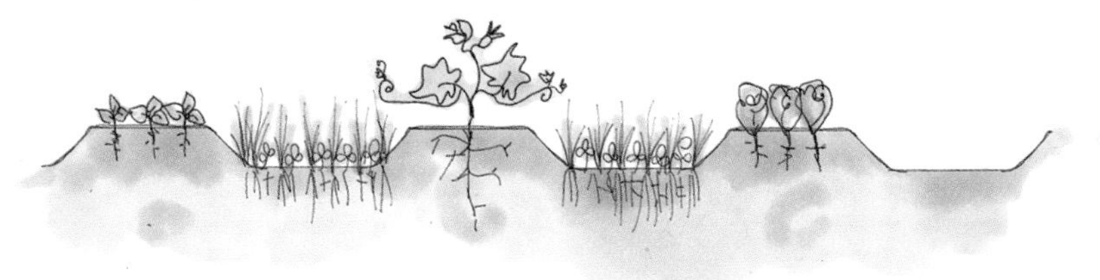

Vegetables grown in companionship with cover crops maximize short seasons to get more cover crops into rotation. Here, spinach and salad will be harvested and cover crop paths mowed, debris will go into paths, and a weed barrier will cover all spaces save the squash bed—allowing the winter squash to spread rapidly across the three beds while compost is made in situ underneath.

FIGURE 27: COMPOST-A-PATH

Bed Height & Shoulder

(3″ to 9″ height, 3″ to 6″ width)

The path receives debris from nearby beds, leaving bed top clean of crop residue for rapid succession planting of new crops! The debris in the path can now be flail mown, top dressed with compost and seeded to a nitrogen-fixing cover crop to balance the carbon-rich debris and create in situ compost. In the future this material is reapplied as finished compost to the bed top to fertilize future crops.

Bed Top

(usually 30″ to 32″)

Bed top width measured from shoulder to shoulder top is usually 30″. Even a 48″ wide bed top is best for bed top mechanized weeding while maintaining raised bed benefits. Note: Gradual shoulders at bed ends are needed for mounting equipment and tractor onto bed top.

When bed top width and path width are the same, then the same equipment can be used to manage both. Ex: flail mower, power harrow, seeders, etc.

Compost-a-Path (usually 30″ to 32″)

Compost-a-Path width varies to suit equipment. Operators might choose a tractor with a narrow wheel base and 26″ tiller to prepare paths for cover crop seeding. In this case a width of 26″ would work, or paths could be even wider, up to as much as 32″ to 36″ to maximize access of carts and allow more compost generation potential with cover crops.

Match Equipment to Architecture

Variation of garden bed building. The Permabed is a raised bed that is never destroyed. Compost-a-Paths are a method for making in situ compost that use a wider path usually 30″. Overall, the architecture of bed tops, paths, and shoulders should suit your equipment. A 30″ power harrow will finish a 30″ bed top nicely, or a 26″ rear-tine tiller can finish the 30″ or 32″ bed top with two passes. A power ridger will power compost a 30″ path perfectly and reform adjacent beds. Match your equipment and bed scales!

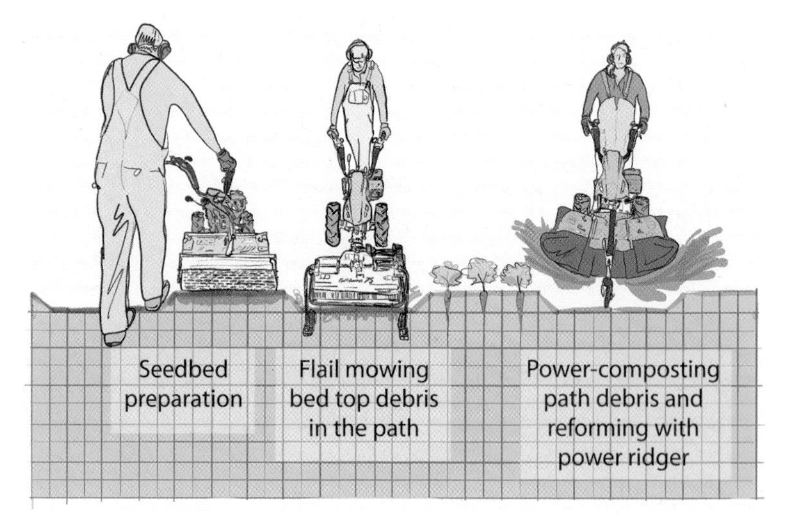

In the Compost-a-Path Method, the same equipment can be used on either bed top or path, depending on the need.

Cover crops in Compost-a-Paths can be spring-, summer-, or fall-sown depending on crop rotation. Fast-growing nursery crops, under-sowing, in situ mulching, and self-seeding are all techniques used for multi-year soil building with cover crops.

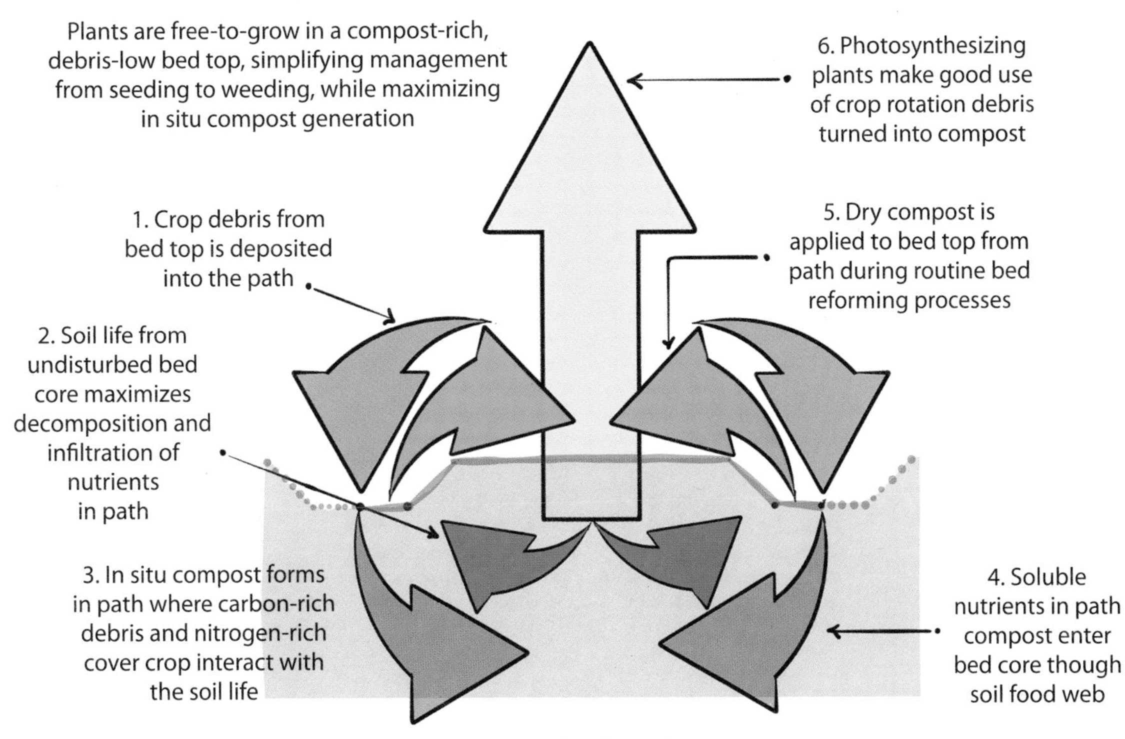

The path receives debris from the nearby bed, leaving the bed top clean for rapid succession planting of new crops and allowing the path to be seeded to nitrogen-fixing cover crops to balance the carbon-rich debris and create in situ compost. Soil life from the core and the ecosystem of the cover-cropped path help to host pest-eating and disease-destroying organisms.

PERMABED SOIL DESIGN

Let's look deep into our soil and understand that we can design our soils by mimicking natural soil formation within an organized garden plot.

DESIGN BOX
PERMABED SOIL HORIZONS

Our garden can form horizons (visible bands within the soil with different characteristics) like natural soils and provide the same soil ecosystem services: air, water, and nutrient storage and release, and habitat for beneficial soil organisms, like mycorrhizal fungi and nitrogen-fixing bacteria. Soil horizons are like sponges for nutrients and water, as well as habitat for soil life that creates rich soil. There is organic matter-rich tilth in the A horizon and nutrient-rich reserves within soil aggregates in the B horizon. The soil life conservation core maintains soil organisms' habitat. Permabeds and Compost-a-Paths are designed to *improve natural soil-forming processes.*

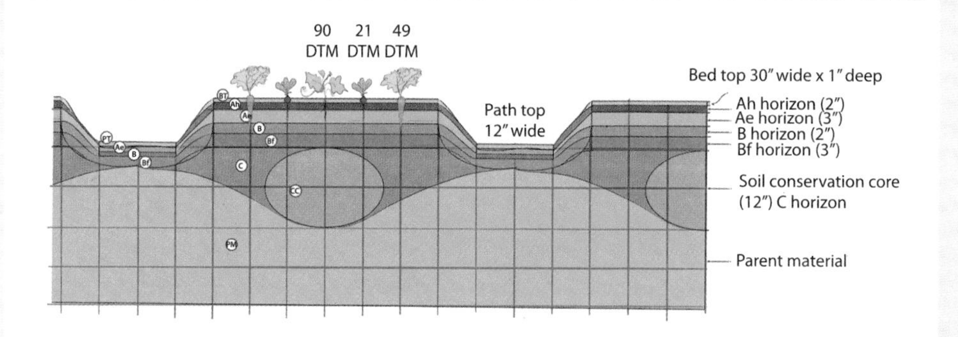

PERMABED SOIL HORIZONS AND CHARACTERISTICS

- **Bed Top (BT):** The layer for compost spreading, crop seeding, light cultivation, crop mulching, and dust mulching.
- **Ah Horizon** (*h, humus layer*)**:** Has the highest percentage of **Soil Organic Matter (SOM),** benefiting from in situ compost and mulch management (power-composting with mower/ tiller/power ridger, etc. as designed for your

equipment guild use); soil organisms are very active in spring and fall. The first crop roots (radicles) grow here in the fine seedbed. The Ah horizon doesn't develop the complex aggregates seen below because seedbed prep limits this. Here, high SOM creates seasonal micro-aggregation, preventing soil cracking, erosion, desiccation, and flooding, all of which kill and stunt young crops.

- **Ae Horizon** (*e, eluviation*)**:** Water percolation pulls nutrients (clays, aluminum, and iron) from the Ae layer and deposits them in the B horizon. Deeper preparation for easy transplanting occurs here using a power harrow that doesn't mix the overlying Ah horizon into this Ae. Crop roots push hungrily through the Ae, connecting to the complex B horizon below.
- **B and Bf Horizons** (*f, ferric*)**:** This horizon is enriched with iron, clays, and other nutrients. Because the B horizons don't get tilled, save for occasional low-impact broad forking or subsoiling (at start-up), they develop a complex aggregation. Complex aggregation bridges into the soil conservation zone—where soil life prospers in relationship with crops but is undisturbed by equipment. Here soil life draws upon the mineral and water wealth of the subsoil, becoming enriched with the sugars and plant materials of the surface world and transferring it to the crops.
- **C Horizon** (*C, Conservation*)**:** The B horizon aggregation *develops downward as porous space is opened and soil structure is improved and stabilized*—nutrients and SOM are transported into and develop the C horizon and its soil conservation core. A well-aggregated C horizon occurs ONLY with best Permabed practices over time. Don't disturb this horizon, let it develop.
- **Path Top (PT):** The zone of human/wheel movement and cover crop seeding. The path horizons form over a three-year cycle, and then compost is applied to the bed top. Flail mowing of debris and cover crops and tilling-in of chopped debris is typical here.
- **Ad Horizon** (*d, debris*)**:** Zone of debris accumulation, pulverization.
- **Ba** (*a, assimilation*)**:** Zone of nutrient and material assimilation into underlying layers.
- **Bc** (*c, compost*)**:** Zone of soil life migration from C horizon to digest and compost material.

THE WONDERFUL SOIL AGGREGATE

The soil **aggregate** is a house for soil life, with macropores and micropores (openings) holding air and water within a mineral (bricks) and SOM (mortar) matrix. Compaction *reduces* porosity. Macropores (> .08mm) hold water when it rains and let it drain freely afterward, allowing air to return to help oxygenate soil for aerobic processes like decomposition. Micropores (< .08mm) store water longer-term for crops. There are even **ultra-micropores** in soils that contain activated carbons!

Overall, aggregates have huge (but microscopic) *surface areas* when the edges and planes of various mineral and organic materials are taken into account; these surfaces provide points for adhesion of water molecules and nutrient ions to soil surfaces and cohesion of molecules to each other within porous soil spaces. Well-aggregated soils have a high *Cation Exchange Capacity* (CEC), meaning they hold and release more nutrients and water than soils with low CEC. Soil life also thrives within these soils and helps regenerate them by connecting the soil's water, nutrient, and communication pathways as it shreds organic matter for decomposition, excretes micro manure, and fixes nitrogen right out of the atmosphere. By building below-the-surface habitat, we regenerate nitrogen-fixing bacteria, mycorrhizal fungi, and other beneficial organisms.

Pro Tip: *In a garden system where waste material is cycled in situ, soil horizons develop and the soil surface has mostly a diverse crop cover in all seasons. This is nature's design. You can see it in a woodland or prairie ecosystem's thriving and resilient productivity.*

EQUIPMENT FEATURE

IMPROVING SOIL

It is important to understand not only how to reduce harm to the soil from equipment (which is what S4-tillage is all about) but also how to use equipment to *improve* soil. Equipment can help improve your soil if it allows more soil-beneficial processes to occur each year.

- **Spreader:** Compost is a great way to build and boost soil health. Soil organisms eat and live in Soil Organic Matter (SOM). SOM helps to cycle and store water and nutrient.
- **Flail Mower:** Efficient cover crop, green manure, and crop debris management adds SOM and nutrients and mulch to the soil.
- **Rear-Tine Tiller** is used to blend compost, crop, and cover crop bits into soil. Shallow introduction of oxygen and debris-rich material allows the soil life to consume and transform it in the A horizon.
- **Power Harrow** doesn't invert the soil and can be used to work deeper into the soil profile without mixing the A and B horizons. This keeps soil healthier for more self-regulation. Even the

prairies had disturbance from bison, and prairies made great soil!

- **Power Ridger** reforms Permabeds with in situ compost in paths. By focusing aggressive actions in paths, we maintain the ecosystem mimicry of a natural soil on the bed top.
- **Roller/Crimper** is used to maximize the leaf litter effect of forests and grasslands with in situ mulch that protects soil from heat/cold temperature extremes, heavy rainfall, snow melt erosion, and compaction from equipment. Plus, the debris blocks weeds!

USING EQUIPMENT FOR CROP ROTATION MANAGEMENT

A substantial part of a grower's focus is on *why, how,* and *when* certain equipment should be used within a **crop rotation**. A crop rotation is a suite of crops and cover crops that follow one another through time: different crops are planted in a given piece of ground each year, one after the other. Sometimes, crops are organized within the bed as crop guilds (three or more crops together), and this is called a **guild crop rotation**. The reason for rotation is best management of disease, pests, fertility, and debris. **Equipment task schedules** help in getting this work done on time. Scheduled tasks might include: reforming beds in fall, mowing cover crops in spring, preparing fine seedbeds in early summer, maintaining bed tops with cultivation, power-composting green manures in early fall, and even chipping orchard-pruned wood into ramial wood chips for paths.

In order to manage guilds in rotation over a season and years. a series of equipment tasks that must be scheduled.

Crop Guild: Salsa guild
Rotation group #1

Crop Guild: Storage roots & garlic
Rotation group #2

Crop Guild: Summer mixed veg
Rotation group #3

FIGURE 28: GENERAL CROP ROTATION

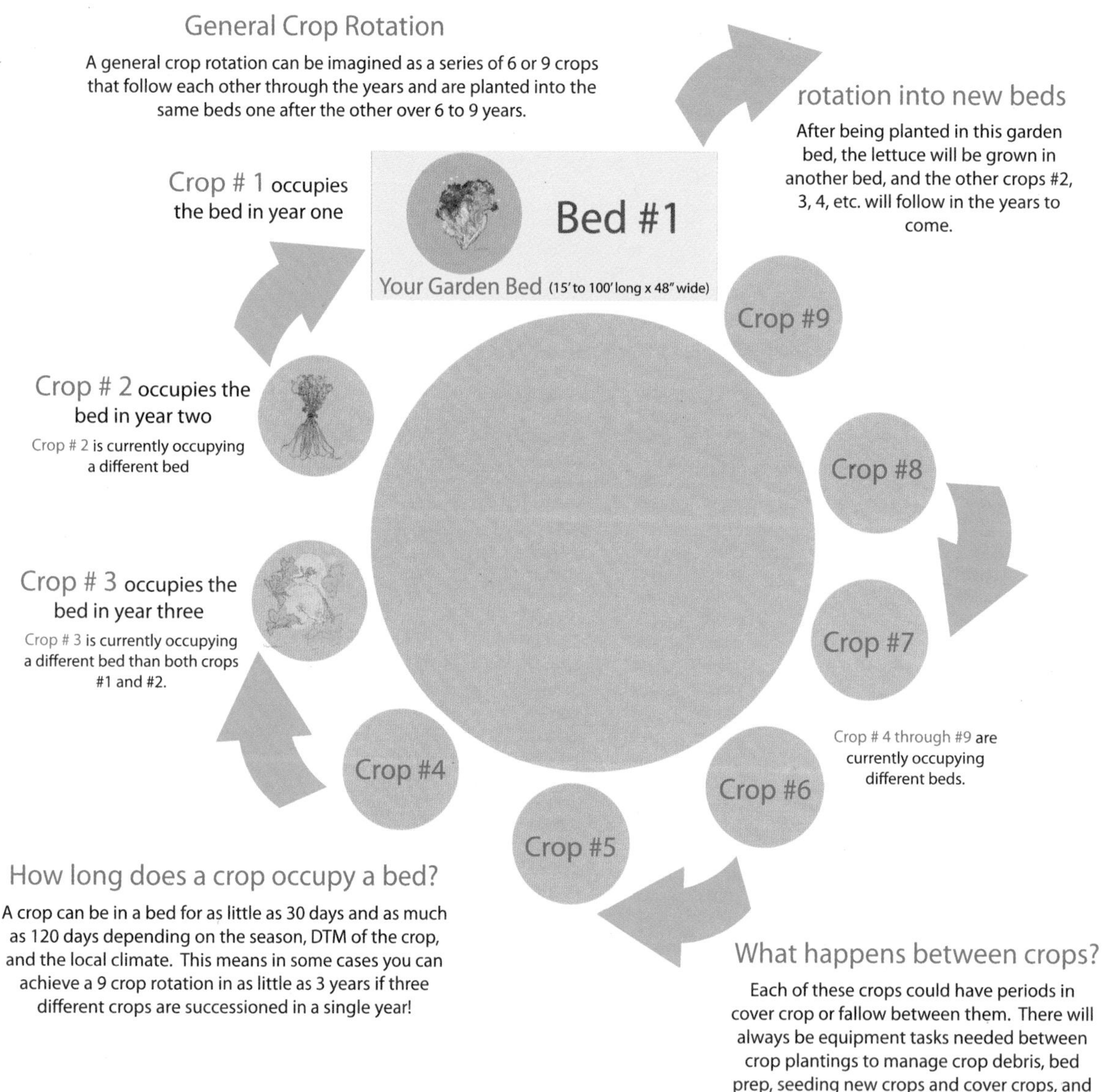

FIGURE 29: SEASONAL EQUIPMENT TASKS OPTIONS AND OPPORTUNITIES

The following figure gives a sense of common tasks and when they might occur seasonally. However, these are options and opportunities which would only be assigned through the process of crop planning and knowing your specific techniques and gardening methods.

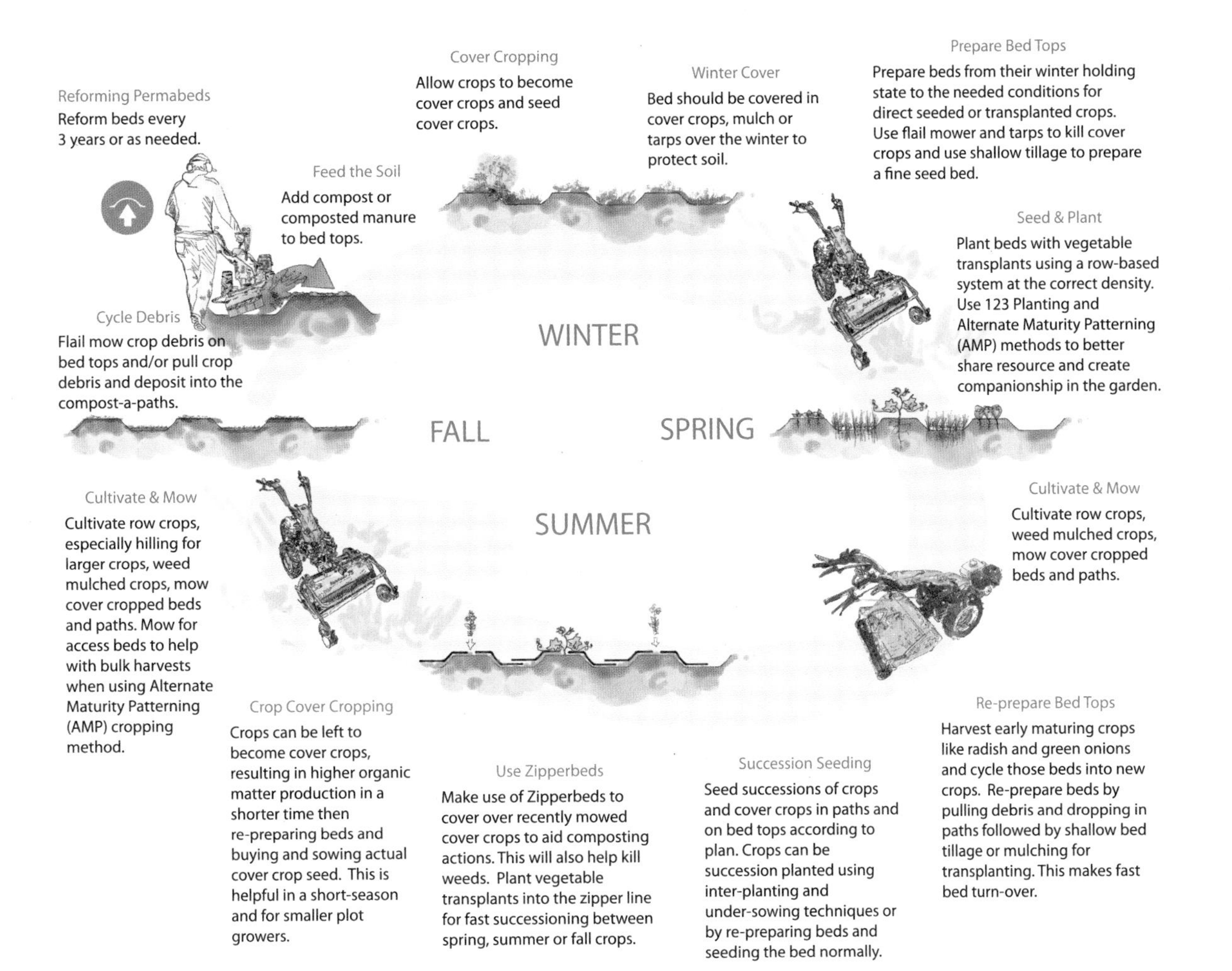

FIGURE 30: COVER CROP PRIORITY

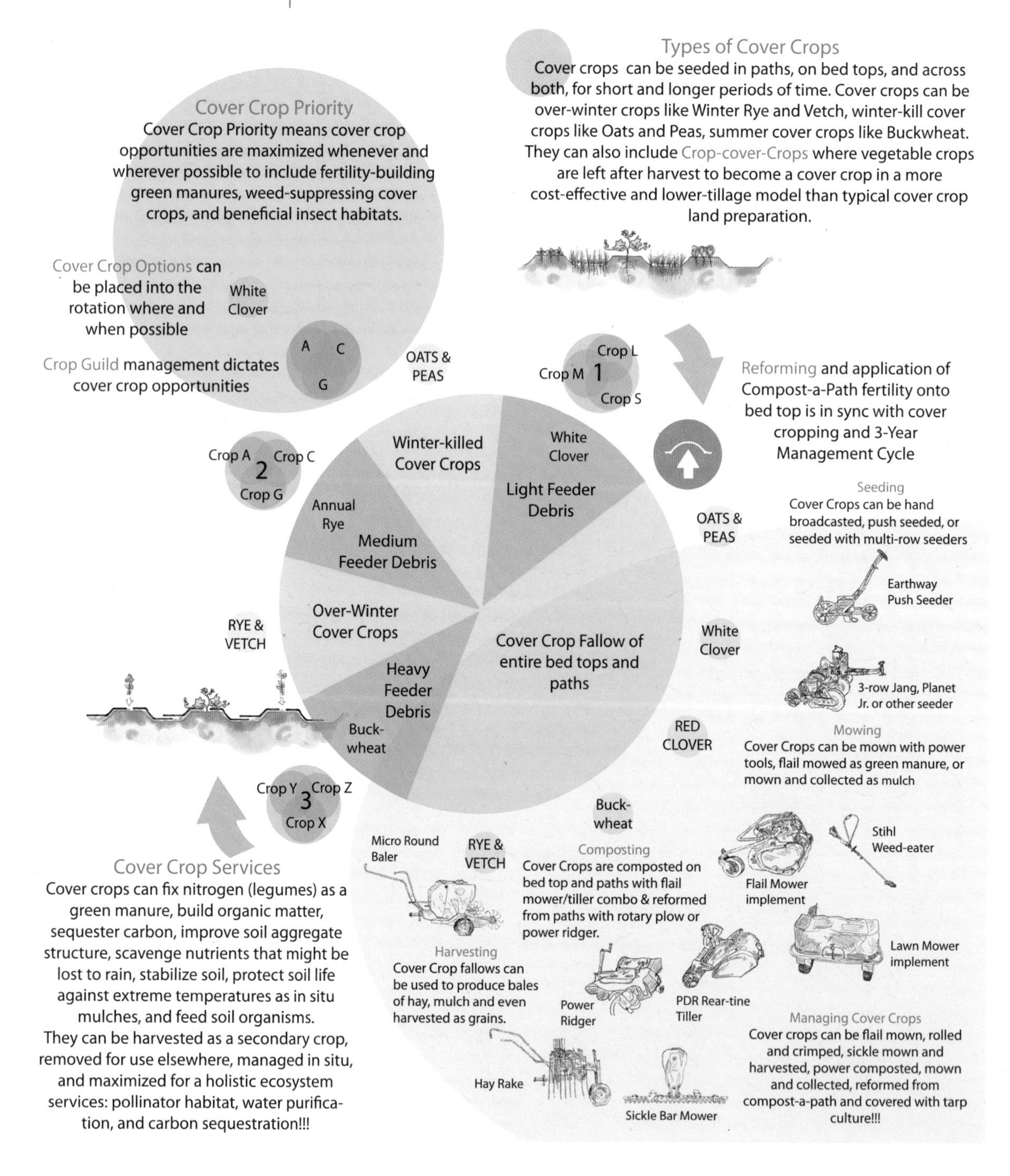

Chapter 7
Growing from Scratch

It is important to understand the full extent of tasks needed to create a Permabed System from scratch: clearing land, forming beds, and seasonal activities. The following pages will go in depth into the processes of turning any piece of land into a Permabed System.

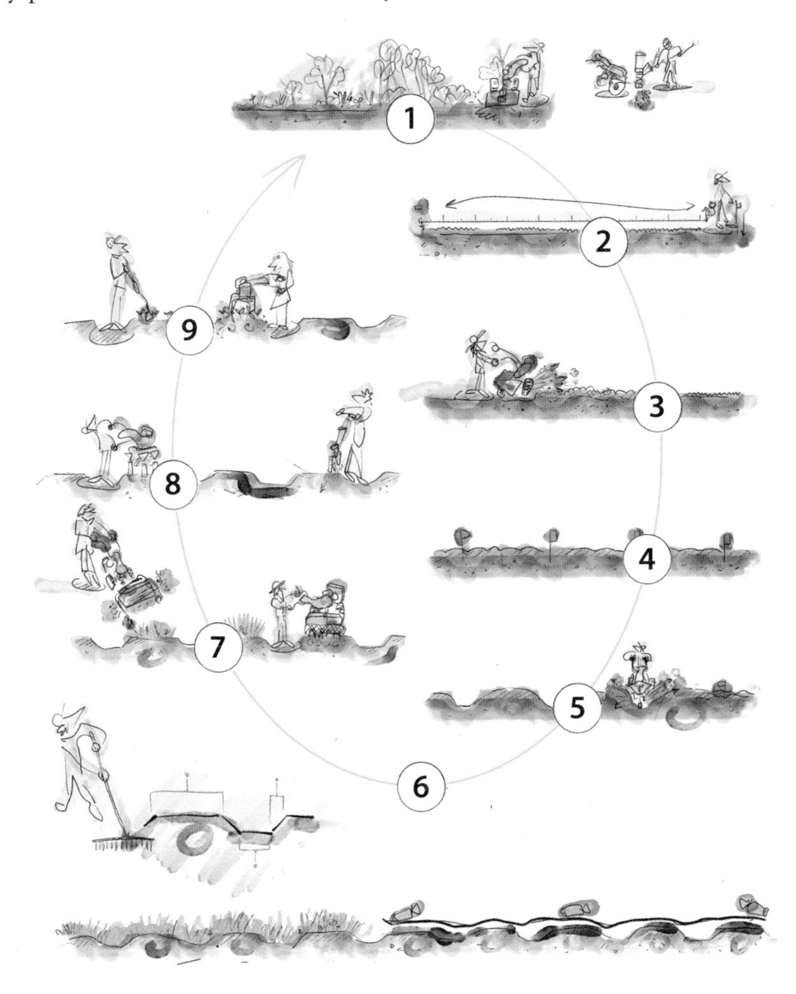

① CLEARING NEW LAND (OR RECLAIMING)

For many growers, the initial work on land entails *reclamation* to create a more organized, accessible, and manageable landscape. Overgrown land (treed or old, weedy fields) can be tamed for future garden, orchard, pasture, or other needs.

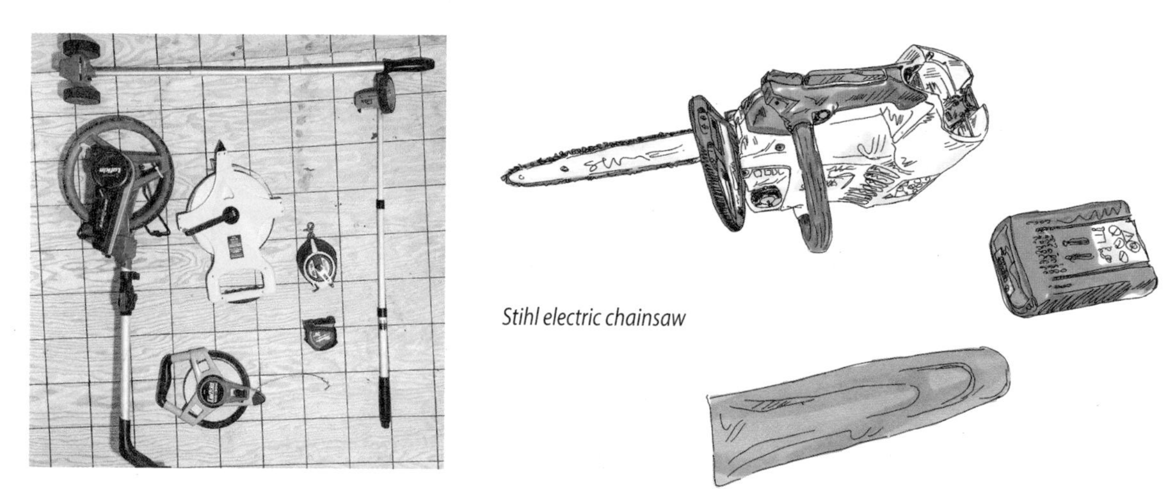

Stihl electric chainsaw

Measuring tools for plot layout

Landscape flags

Poly fence posts

FIGURE 31: CLEARING LAND AND RECLAIMING A FIELD

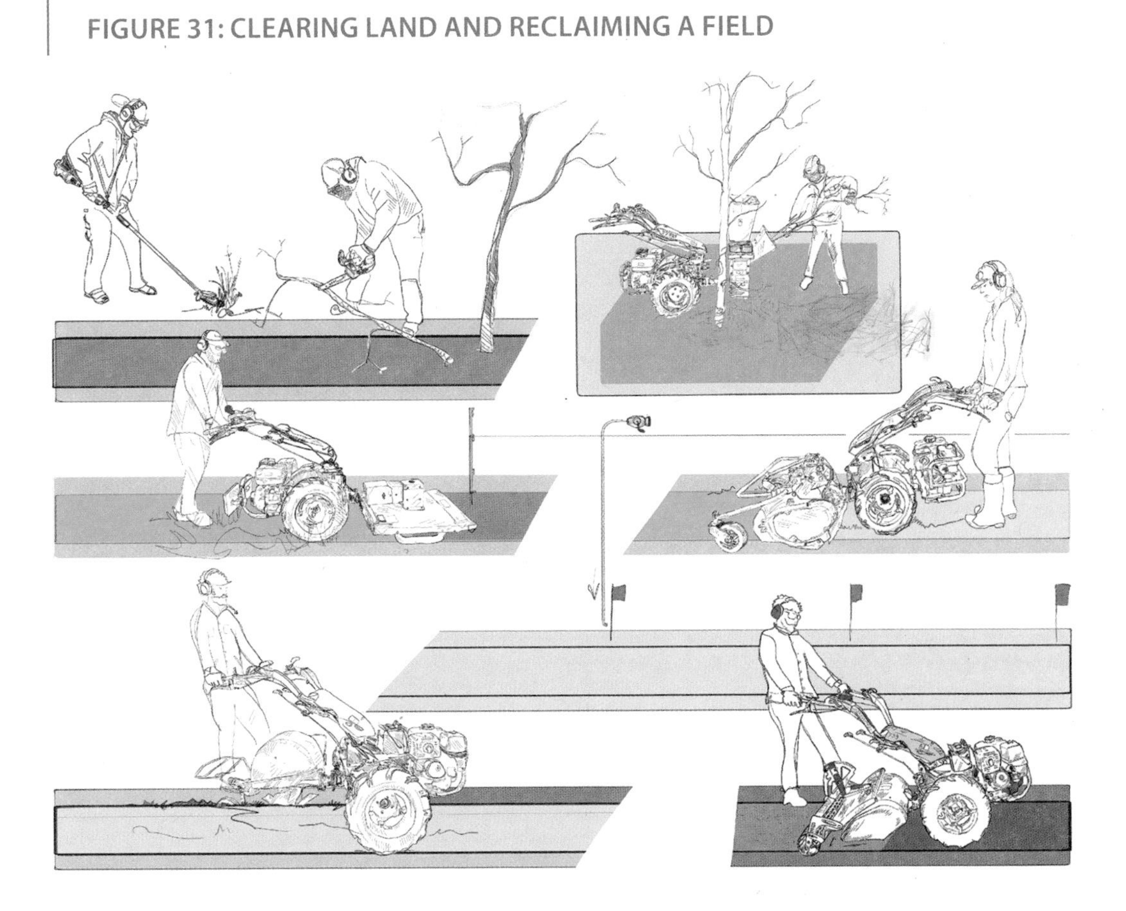

- ***Measure, mark, and lay out the project area*** *using a rolling measuring wheel (Lufkin), measuring tapes, marker flags, and poly stakes.*
- ***Cut down large shrubs and trees*** *using a chainsaw. Small electric chainsaws are great for most stems and limbs (1" to 6") and are easy to operate, quiet to use, fast to charge, and affordable to buy.*
- ***Chip woody material*** *using a chipper/shredder; the chipped material can be applied as mulch for paths later.*
- ***Remove stumps.*** *Plowing can cut through small annual and perennial roots. However, larger trees will need to be pulled out with a backhoe.*
- ***Mow large weeds/shrubs*** *using a brush mower or a flail mower. This will bring the level of the greenery down enough to allow easy operation of plowing equipment.*
- ***Finish mowing*** *using a flail mower—or a lawn mower, if you have one. This will make plowing easier.*
- ***Measure, mark, and lay out with measuring tape and flags for plowing a Permabed*** *by organizing the fields according the intended plot and bed widths, lengths, and quantity.*
- ***Plow field*** *using rotary plow or swivel rotary plow.*
- ***Prepare fine seedbeds*** *with PDR tiller or power harrow.*

② GARDEN PLOT LAYOUT

Now that the land has been cleared, you can lay out new garden plots. Measuring out the plots can be done with some poly fence posts (they are portable and have spikes that are good for placement in different soils), along with some stronger markers that can be pounded in for accuracy (5⁄8" rebar or 1.5" wooden stakes). Most sites will require a 200 ft or 400 ft tape measure. A hand tape is also useful for measurements between bed paths later and checking that the plot is square.

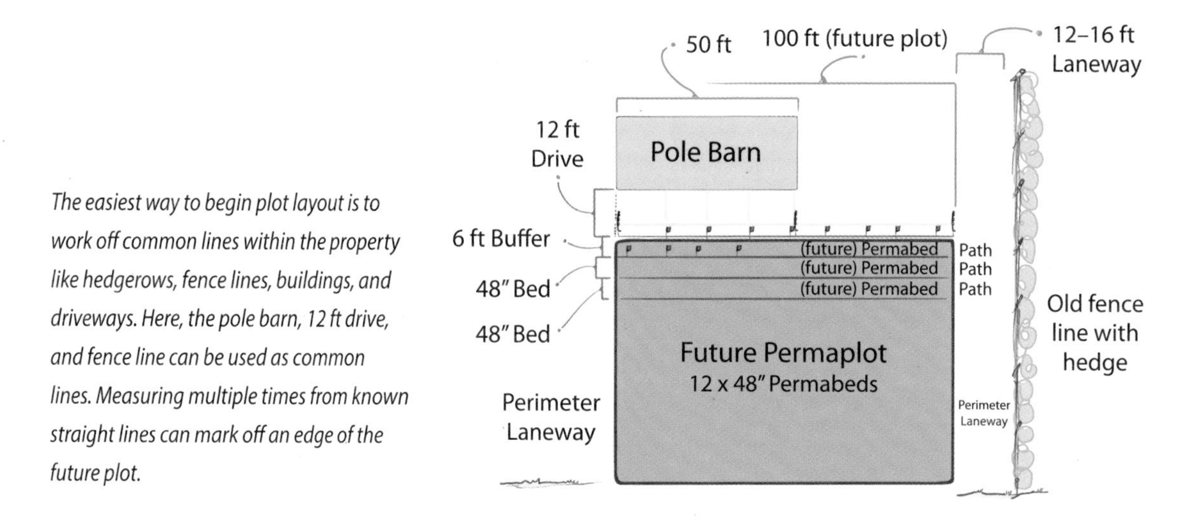

The easiest way to begin plot layout is to work off common lines within the property like hedgerows, fence lines, buildings, and driveways. Here, the pole barn, 12 ft drive, and fence line can be used as common lines. Measuring multiple times from known straight lines can mark off an edge of the future plot.

Measure from a common line—a building, fence, or drive—and mark the straight line with flags. **Example:** Your common lines include an old pole barn, a 12 ft laneway, and a perpendicular fence. You want an additional 6 feet of grassy perimeter to mow and maintain between the lane and the garden. Measure multiple times along the building's length, and flag the future garden's edge every 6 feet at 12 ft and 18 ft points. Pound a stake at the corner and run a measuring tape following the flags to full plot length (100 feet); even if the building is only 50 feet, you would have enough flags to establish a straight line parallel to the building for the entire 100 ft field by eyeing down the line. You can now proceed with actually marking off and squaring your entire plot.

HOW BIG IS THE PLOT?

Put a post at each of the four corners according to the desired size. Actually, put posts 6 feet wider (your buffer) all around because stakes and string get in the way of plowing. The 6-foot perimeter buffer can be seeded into a ground cover (a bee mix or micro lawn).

Plot width = bed QUANTITY × bed WIDTH

Plot length should fit your field size and use a standard bed length (see Figure 25 in Chapter 6).

Example: A Permaplot with 12 beds at 100 feet would be 48 ft × 100 ft (+ perimeter buffer).

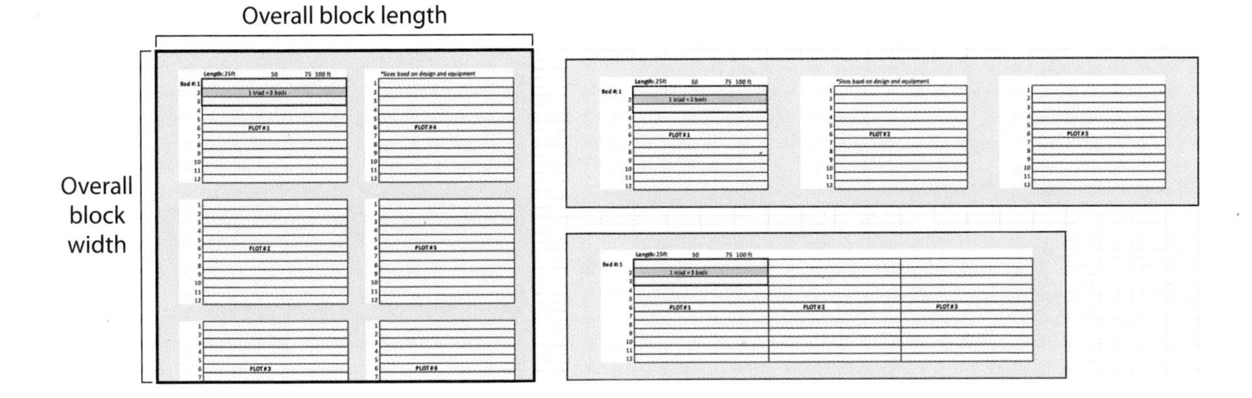

Different layout strategies for plots as blocks include having laneways between plots, as well as more linear or square layouts.

DESIGN BOX

A NERDY WAY TO SQUARE YOUR PLOTS AND BLOCKS

Roughly squaring a field can be done by plowing parallel to a fence line. But small-scale growers often subdivide larger fields. Plots should be square, meaning there should be a 90-degree angle where plot length and width meet. If not, you'll be trying to fit equal-width beds into a plot plowed askew. **The Pythagorean theorem** says this: *the sum of the squares of the legs of a right triangle is equal to the square of the length of the hypotenuse: ($a^2 + b^2 = c^2$).* Using this math can square your plot! The hypothenuse of a **48 ft x 100 ft plot** (12 x 4 ft Permabeds x 100 ft long) is *110.92 feet*. So, if you measure from stake #1 to stake #3, it should be 110.92 feet *(if square)*. If not, then adjust stake #3. You can also use the **3-4-5 technique for squaring** to check the angle at the corner of the new plot (stake #2). If you measure 3 feet from a point and then 4 feet at a right angle to that line, then the hypotenuse is 5 feet. That's a short form of the old Greek dude's theorem.

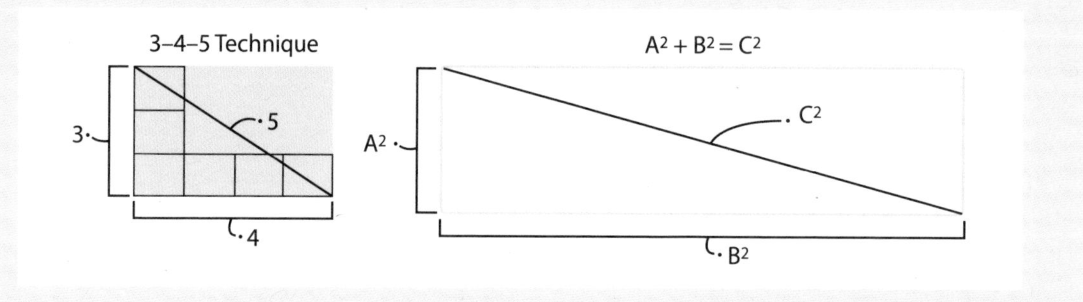

3 MICRO-PLOWING A FIELD

The land can now be plowed using a moldboard, rotary, or swivel plow. The moldboard plow is the cheapest option, but the rotary plow is more versatile in different soil conditions. The swivel plow is best for large fields and sloped ground. (See Equipment Feature in Chapter 3, "Three Reasons to Use a Swivel Rotary Plow.")

PLOWING A MID-LINE TRENCH AND FORMING YOUR MID-RIDGE

This example uses the rotary plow to "micro-plow" an old field. First, make sure the field is marked off and squared, so you don't plow land unnecessarily. Then plow along a flagged line that is *centered in the plot's mid-line.* This **first pass** follows the flagged **mid-line** of the new plot. The **second pass** continues to widen this trench in a returning pass. A **third pass** is advisable to make the trench even wider. Now, fill in the newly made trench with passes four, five, and six to fill the trench and form a **mid-ridge. Remember,** the plow only jettisons soil in one direction, so trenching passes require your left tire to be in the trench *(save for the first pass that makes the trench)*, and filling passes require your right tire to be in or near the trench *(or to the left of the new mid-ridge)*.

Why refill the plot's mid-line trench? Because you don't want any unplowed ground left in the middle of your new plot! The first three passes are trenching into unplowed sod and packed soil. If you just start making concentric plowing passes forming your mid-ridge without first making trenching passes to form your mid-line trench, then the soil underneath the plot's center mid-line will *never* be plowed. In other words, the *initial two or three passes* make a trench and fill it, so when subsequent passes mound soil over that mid-line to become a **mid-ridge** (Figures 32 and 33), the ground beneath the ridge is effectively plowed too—not just piled on top of unplowed sod. This new mid-ridge becomes the central line around which all further plowing passes are made.

Pro Tip: *The rotary plow has a baffle plate that can be added/removed. Removal of the plate allows more soil to be thrown—and thrown further. Depending on soil conditions, you may choose to use or remove this when plowing and/or bed forming/reforming.*

PLOWING YOUR FIELD

The mid-ridge is the focal line for a rectangular and clockwise plowing pattern around either side of the mid-ridge. Keep your right tire in either of the trenches; your rotary plow will dig into unplowed soil on ridge-left and jettison more soil onto the widening mid-ridge that is becoming your **plowed plot.** Make continuous passes on each side of the mid-ridge, growing the plowed space between with each pass (5, 6, 7, 8, 9, 10, etc.)

FIGURE 32: PLOWING THE MID-LINE AND FILLING AND FORMING A MID-RIDGE

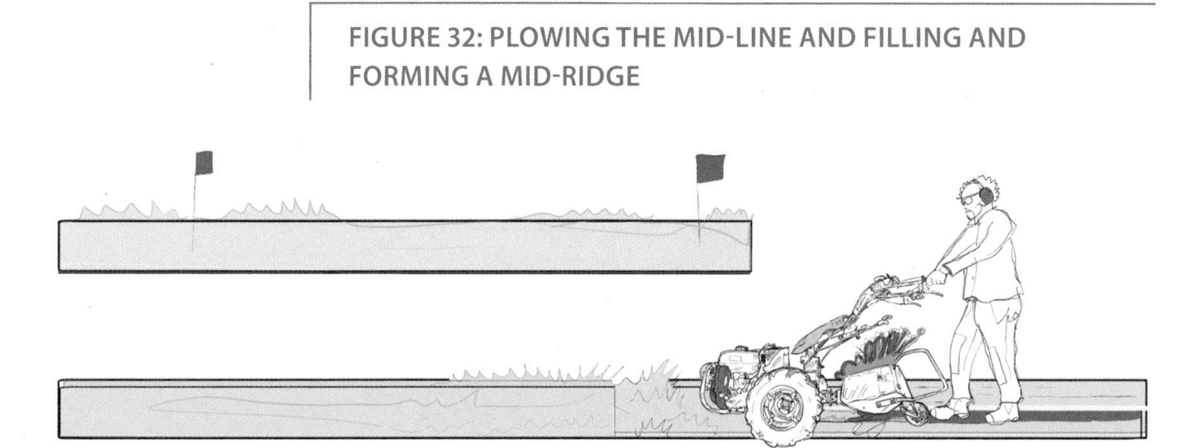

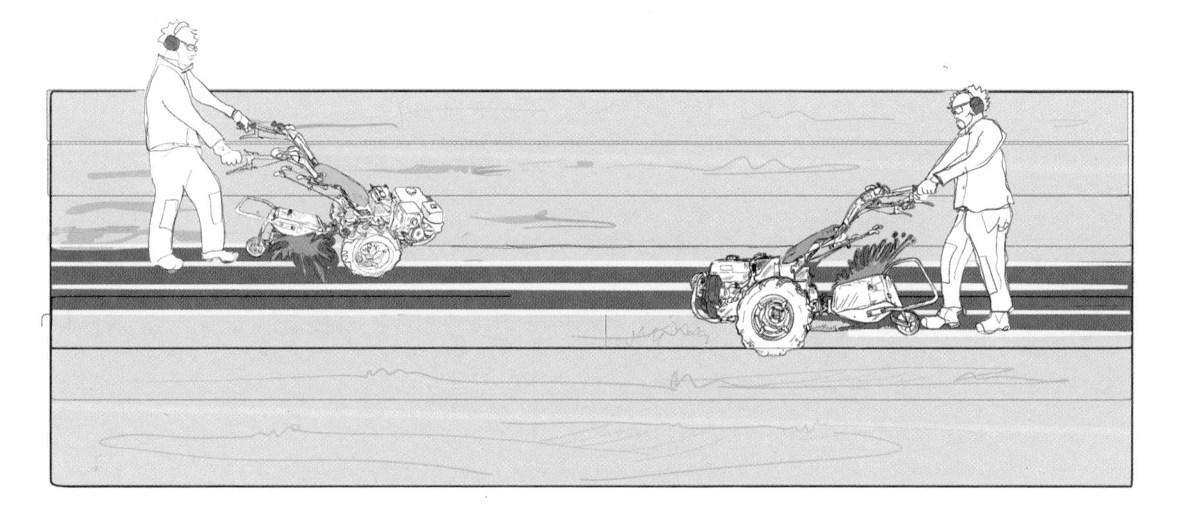

Top of image (with red flags): *Plot layout should include staking all the corners and the plot center and marking the mid-line with flags.*

Middle of image (with one plow person): *The single yellow line indicates the first of the three passes needed to form a deep* ***mid-line trench****.*

Bottom of image (with two plow people): *The yellow lines indicate the filling passes made with the rotary plow turning soil to the right to form the* ***mid-ridge****; after that, subsequent passes continue plowing in a clockwise pattern around the ever-widening plowed plot.*

FIGURE 33: CUT-AWAY VIEW OF A MICRO-PLOWING

Layout for Plowing

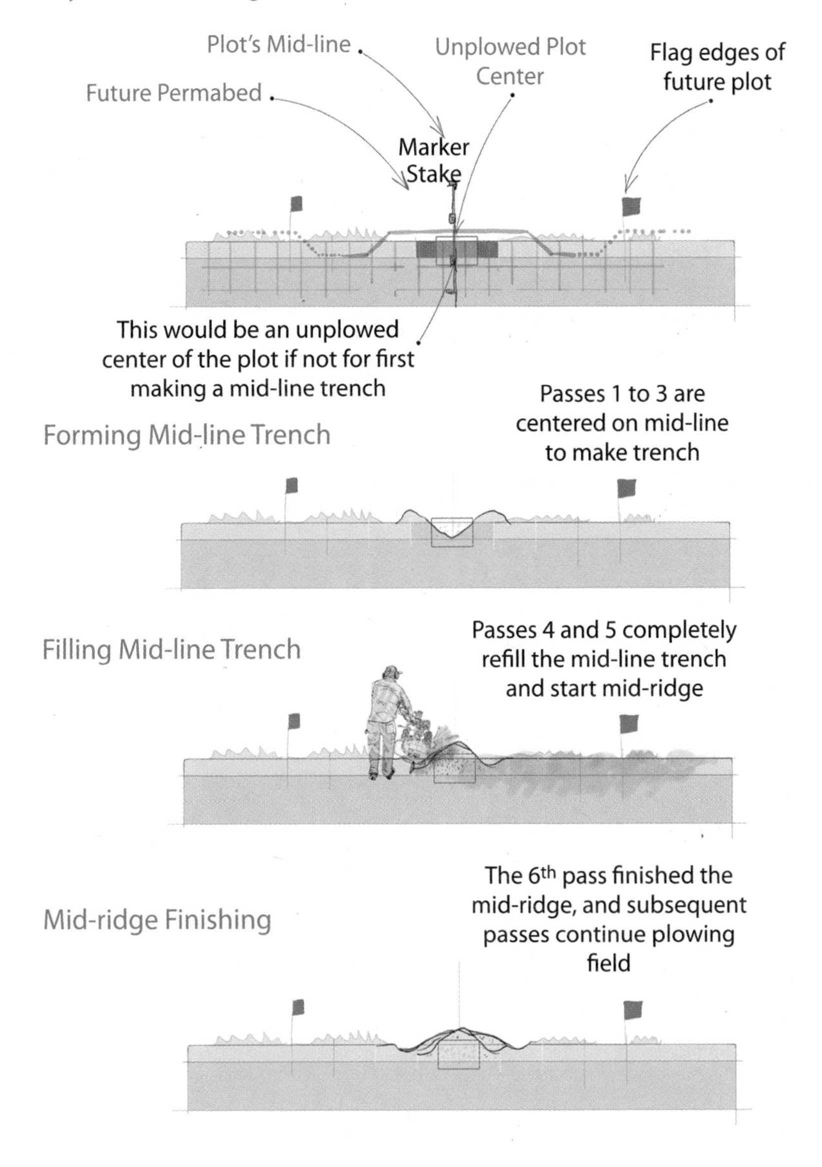

FIGURE 34: NEW FIELD TO FINE SEEDBED

Sometimes, Row Crops Farmers (and other growers) don't want to use raised beds. Here is the process of preparing ground for row crop farming.

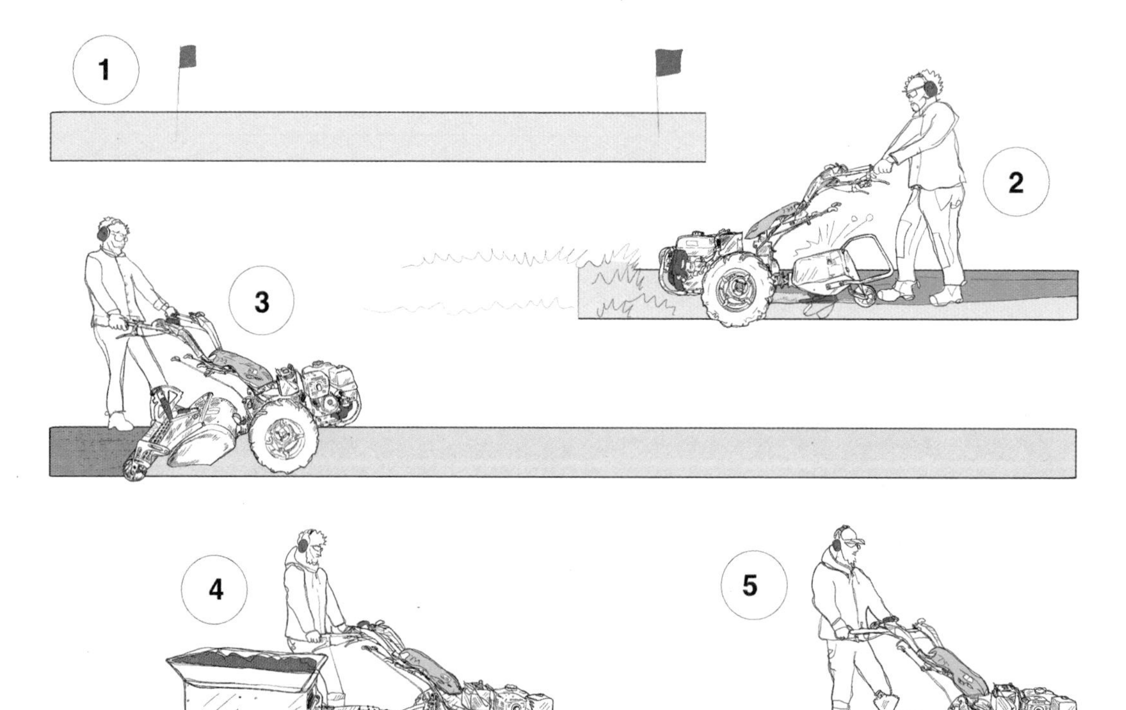

1. ***Lay out*** *the new garden plot.*
2. ***Plow*** *the entire plot with any plow type.*
3. ***Remove plow ridges*** *using a rototiller.*
4. ***Fertilize the garden*** *using a compost spreader.*
5. ***Finish seedbeds*** *using a power harrow, mixing in compost and packing the top for good soil-to-seed contact.*
6. ***Now you can mark rows and seed,*** *then use your row crop tractor for cultivation.*

4 PERMABED LAYOUT AND BED BUILDING OPTIONS

Now that the field is plowed, you can pass with the rear-tine tiller to break up the plow ridges. Then, flag the plot for proper Permabed layout. Optionally, Row Crop Farmers can leave this flat (see Figure 34) and go right to cover cropping or preparing for a fallow period with S-tine cultivation to remove weeds.

BUILDING PERMABEDS

There are multiple methods for building Permabeds, and layout depends on which *approach* and which *implements* are used. Some of these methods require plowing the ground and *then* building beds; other methods do bed forming in conjunction with plowing.

Pro Tip: *Overall, the rotary plow is the best equipment for opening new land, but when I am working on larger fields, the power ridger is my preference for building Permabeds and reforming. However, the tiller and hiller/furrower combo served me well for many years as I was starting up, many years ago.*

DESIGN BOX

LAYOUT AND BED FORMING

THREE WAYS TO MARK LAND

1. If the land is already tilled, you can mark it by measuring with two stakes and a string and making marks with your boot into the fresh soil at the measured distance (bed width). The correct distance is set by making sure the length of string between the two stakes is the width of one Permabed. One stake is placed at the edge of the plot, and the other end is placed (with a taut string) 48" into the plot to mark the path. This mark designates the line to follow with the power ridger or furrower to form a new path. This process can be continued up the plot to mark off the path in several points to be followed with the tractor and power ridger, and then this

new path can be used as a point of measurement to mark off the next path in the adjacent bed that will have its path furrowed next.

2. If working from rough plowed land that hasn't been tilled (which is feasible, but less recommended), you can use landscape flags to mark out the path center and then follow that line with a power ridger or hiller/furrower.
3. Lastly, when using a rotary plow, you need to mark the mid-line of the bed, *not the path*, and form around it by emptying a trench and refilling and forming the bed as discussed below.

MANY WAYS TO MAKE RAISED BEDS

1. **TILLER AND HILLER/FURROWER OR POWER RIDGER:** *The land must first be plowed and tilled.* **To Form Beds:** Mark *all future paths* with flags (or using two stakes and string) and drive the two-wheel tractor down the middle of the future path and move soil out of the path, applying it onto the two adjacent beds. A helper can pull the flags out ahead of the tractor as the operator keeps equipment straight by eyeing down the flagged line. Then prepare and finish the bed top with a power harrow or tiller. The hiller/furrower combo requires multiple passes and some raking to make a wider Compost-a-Path. The power ridger operated down marked lines that are 60" apart will leave a 34" trench/path with one pass (in previously plowed and tilled ground) and a 26" bed top. After a pass with a PDR tiller, the bed top will widen to 30", and the Compost-a-Path will narrow to 30". ***Note:*** The furrower or power ridger will only be displacing soil to one side if secondary *path-widening* passes are done.
2. **ROTARY PLOW:** The land should be plowed and tilled for this method, or you can build the beds as you plow and then till afterward to make the flat bed top. **To Form Beds:** Mark the mid-line for each future Permabed and make sure each line is the same distance apart as your beds are wide (48"). Follow plowing procedures to make a wide trench centered on the mid-line with ***passes 1, 2, and 3.*** Remember, the plow only throws soil to the right! ***Passes 4 and 5*** will fill the trench and form a mid-ridge. ***Pass 6*** will finish raising the bed mound. Subsequent passes can be used to widen the raised bed, or you can continue like this with your right tire always in the trench to plow the entire field in a clockwise pattern of ever-widening passes around your initial mid-ridge. Use a rear-tine tiler or power harrow to finish the bed top.

Note: If your goal is a wider raised bed with good depth of soil for easy planting, then it is essential to first make a wider trench using more passes to throw soil out of the trench (future bed center) and then fill and build it back up.

3. **BED SHAPER:** *The land must be plowed and well-tilled.* **To Form Beds:** Mark paths and straddle each future bed and pull the soil into the center with two discs.
4. **HAND TOOLS:** *The land should be plowed and tilled.* **To Form Beds:** Bed can be made with shovels, hoes, rakes, and other hand tools. ***Note:*** These are often employed along with tractors to refine bed lines.
5. **FOR LANDSCAPERS:** *The land doesn't need to be pre-plowed for this method.* **To Form Beds:** The rotary plow method can be used to make perennial beds with grass lawn strips between them by simply marking off the gaps between the beds and staking future bed center at the mid-line (A), mowing the area on either side of the stakes and plowing the bed space to form beds using the direction for forming beds with a rotary plow (B), and then tilling and/or power harrowing the bed top (C). The final product is a series of Permabeds with undisturbed sod between them, making it easy to access the beds and prepare the bed tops using a two-wheel tractor (D) (See image).

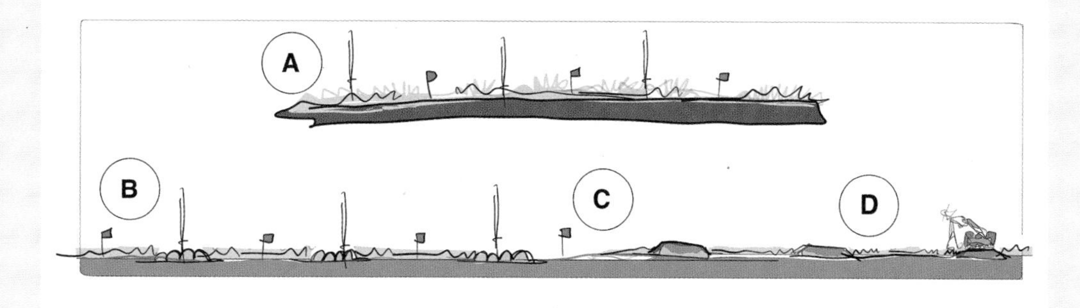

FIGURE 35: HOW TO PLOW A FIELD AND USE A ROTARY PLOW TO FORM PERMABEDS

1st Pass

Follow mid-line to start the trench!

Get Started...

Measure the plot, stake edges and plot center at each end, then flag or mark the mid-line between center stakes.

2nd Pass

Place left tire in the trench and widen the trench along mid-line.

3rd Pass

Keep left tire in trench to finish widening the trench.

4th Pass

Start filling trench with right tire in trench!

5th Pass

Keep right tire in trench to continue filling and mounding.

6th Pass

The last pass has formed the mid-ridge.

Keep Going!

A mid-ridge can be widened to form a Permabed, or it can be the focal line for continuing passes to plow an entire plot!

Note: It is best to plow an entire field, pass with a rear-tine tiller, then follow bed-forming passes for more depth and uniformity of beds. Only plow when dry!

Pro Tip: *You create a wide trench by operating the rotary plow with multiple passes in a counterclockwise motion. Then you build a bed by operating the rotary plow in a series of clockwise passes to fill the trench.*

FIGURE 36: MANY WAYS TO FORM PERMABEDS

There are many ways to form a bed—the question is: which is scale-suitable and enterprise-specific?

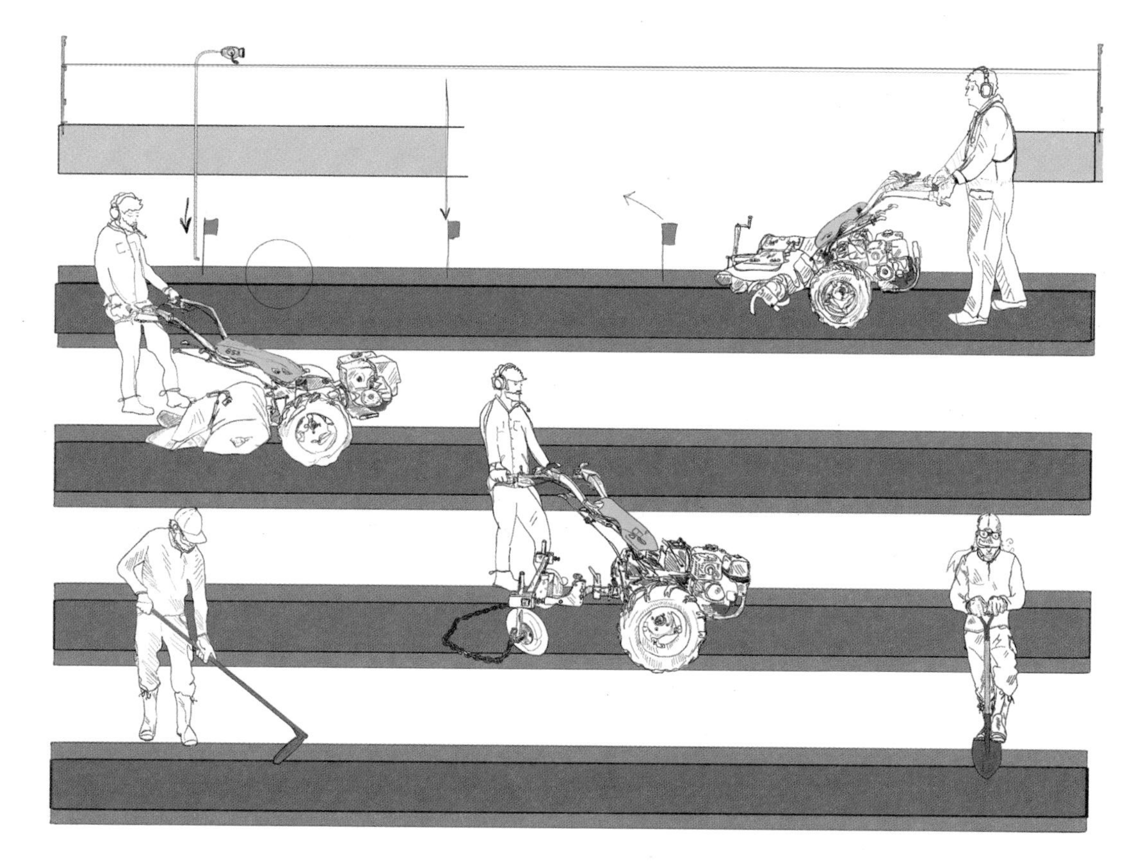

1. **Measure and lay** out paths or bed centers.
2. **Power ridger forms beds** by emptying paths onto bed tops.
3. **Rear-tine tiller with hiller/furrower** is a low-cost way of forming beds by furrowing out the paths loosened by tilling.
4. **Bed shaper makes a low bed** with one pass in well-tilled soil.
5. **Shovels and grub hoes** can also do the trick for plots under .5 acres—when there are many hands. Also useful for cleaning up tractor bed-forming work.

5 BUILDING BEDS WITH A POWER RIDGER

Because it is so straightforward and efficient, let's look in more detail at this method of building Permabeds. This method relies on the power ridger, which has many other future applications. ***Note:*** It is important to remember that this bed-forming time is well worth it; you only have to do it once since Permabeds are never destroyed!

1. **The land must be previously worked:** Plowing and tilling are recommended, but at a minimum, plowing is needed.
2. **Layout review:** Use the lines made in your plot layout to mark and flag paths for bed forming. Make sure to leave your 6 ft perimeter and mark paths 48" apart for normal Permabed architecture, and 60" or 66" apart for the Compost-a-Path Method.
3. **Mark your perimeter line:** Tie a taut string between corner stakes along a perimeter line that is parallel to your garden beds. ***Note:*** *The stake-and-string method* is a 1- or 2-person job.
4. **Mark your first path:** This can be done by measuring and placing flags or marking them into the previously tilled soil with your boot, or even spray paint. The flag method is great if you want to lay out a whole plot in one phase and then build all the beds later. Otherwise, the other methods work well if you just want to mark the soil directly for each new path before making it. ***Note:*** *Flagging* is a 2- or 3-person job.
5. **Flag method:** Person #1 walks along the perimeter string line, and person #2 walks in the field holding the tape measure at the 6 ft mark. Person #3 is flagging the first path at exactly the 6 ft measurement. Walk and flag a point 6 feet from the perimeter line every 5 to 10 feet.
6. **Continue marking paths:** The 3-person team continues measuring and flagging across the plot using each row of flags as the new line for measuring the 48" bed to mark the middle of the next path. ***Note:*** Place flags straight and measure from the flag bottom (there is nothing exact about the top of a flag).

REMEMBER: You can have *any* chosen bed width, but a *bed's width = center-of-path to center-of-path*. A 48" Permabed has these measurements: a (30" bed top) + (2 × 3" shoulders = 6" of total shoulder) + (12" paths [2 × ½-path width is 12"]), so (30" + 6" + 12" = 48").

Note: A 66" Permabed has a 30" Compost-a-Path, so width would be: (30" + 6" + 30" = 66").

7. **Making the paths and beds:** Person #1 operates a tractor with a power ridger. Line the tractor up centered on the flagged line or soil markings. Lock the differential and operate in 1st gear with ⅔ throttle for the most control. *Only do this when the soil is dry.*

   Then, the tractor operator trenches out the path—moving forward and keeping an eye on the line of flags ahead. This is actually easier than it sounds—as long as when you have the tractor centered on a marked row. The soil of the future path will be jettisoned onto the adjacent bed tops. The power ridger leaves a 34" trench that is reduced to 30" when shoulders are formed by passing the power harrow or PDR tiller on the bed top.

   For the flag method, person #2 will simply walk ahead of the tractor operator and pull the flags as the tractor approaches. As soon as a flag is pulled, the tractor operator can proceed to use the *next flag* as a point of reference.
8. **Now, your Permabeds are built.** Or, they are at least roughed in; there is work yet to make them perfectly formed for the long-term and to create finished bed tops for easy seeding, weeding, and harvest.

6 REFINING YOUR ROUGH PERMABEDS

Once the beds have been formed, the bed top can be cross-raked to tidy up the shoulders. The bed will still have a lot of debris *on* it and *in* it that will need to decompose.

REMOVE PERENNIAL WEEDS

Tilling and preparing a seedbed at this stage would result in a weedy mess—especially if perennial grasses or thistles are present; they would only just be chopped into small pieces, each of which would multiply! Remove perennial weeds with one or more of these methods: heavy cultivation with S-tines, digging them up and hand pulling, or using tarp culture.

TARP CULTURE

Use **tarp culture** to get rid of perennial weeds. You can do this before plowing, after plowing, or after forming rough beds. Cover the ground with a poly tarp (most effective) or heavy-duty weed barrier (works also) to fry weeds and decompose debris rapidly. Analyze your weeds. The worst weeds will need 6–9 sunny months to be destroyed, but the least aggressive can be suppressed in 1 month. This is why it is best to build a garden well ahead of planting time. Skipping this stage means you will have to fight the weeds later—among the carrots.

There are many benefits to using tarp culture on beds: tarps get held down in paths more effectively against wind, uniform pieces fit to triads can be used again for pre-weeding in the future, and it is easier to get going with final bed prep once the tarp has worked its magic.

On the left, weed barriers cover new Permabeds for a permaculture market garden in Quebec. In center, poly tarps pre-weed for tree crops on Loeks Farm. On the right, cover crops planted on Permabeds provide winter soil protection for future orchard crops at the Edible Biodiversity Conservation Area.

COVER CROPPING ROUGH PERMABEDS

Cover crop immediately after removing the tarp, or before the tarp even goes on (if you don't have *aggressive* weeds). The benefit of cover cropping beds before putting them into active garden services is four-fold; cover crops will: 1) scavenge excess nutrients from decomposing debris and store it for slow-release fertility, 2) keep soil healthy until you are ready for active gardening (a good placeholder), 3) stabilize newly formed beds with root networks, and 4) contribute their mown organic matter to form aggregates.

7 SEASONAL BED PREPARATION OPERATIONS

It is time to finish Permabeds for active garden service. These next tasks are similar to normal seasonal garden operations for re-preparing beds each year. First, the cover crop can be removed by several methods: another round of tarp culture for four weeks, flail mowing and shallow tillage, or flail mowing and tarp culture, or turning the cover crop into in situ mulch. If you are not using in situ mulching, then finish the seedbed using a PDR tiller, power harrow, or even just a rake.

Pro Tip: *Low-debris bed tops are required for most direct-seeded vegetable crops. For this, flail mowing debris, burying the debris with soil, and short-term tarp culture are the most effective means of providing a weed-free and debris-free seedbed with good tilth.*

8 SUCCESSION PLANTING AND CROP GUILD DESIGN

Succession planting of different crops provides a continuous supply for spring, summer, and fall. Seed short-DTM crops (radishes) every 7 days in spring, and medium-DTM crops (carrots) every 2 weeks in spring/summer; longer-DTM crops (squash) are planted once. The short-DTM crops (spinach or cilantro) can be intercropped with longer-DTM crops (melons) using **Alternate Maturity Patterning (AMP)**, where adjacent beds always have crops with different DTM. See two crop guild variations of AMP in the images here.

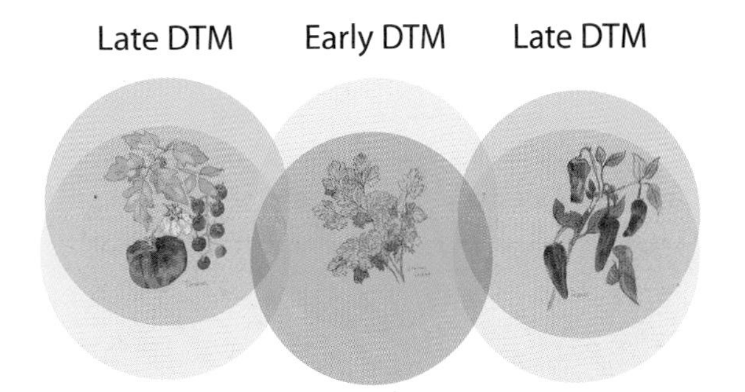

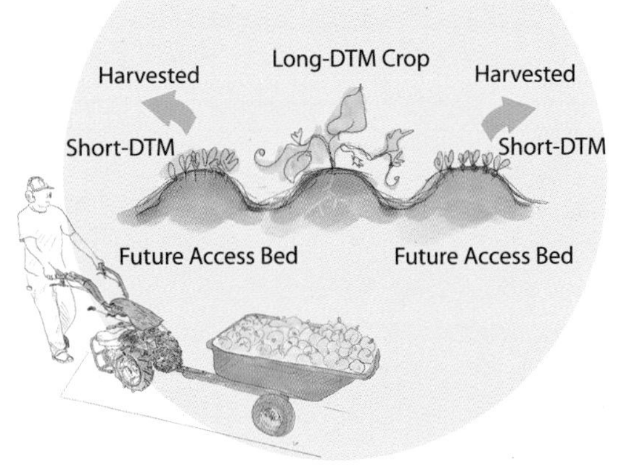

This pattern improves accessibility for two-wheel tractors because when short-DTM beds are long past harvest (and now growing cover crops), long-DTM crops are ready for bulk harvest. Tractor and cart can easily drive up an adjacent bed in clover to help bulk harvest. The Compost-a-Path Method also provides generous access for trailers and carts.

Pro Tip: *Seeding successions can be done with a tractor-pulled seeder or push seeders. Pro-up growers will often have a few seeders calibrated to popular succession crops. Why? Because some crops have unique seed sizes and/or low germination, but are very valuable if done right—like golden beets. If this crop is to be seeded routinely in spring, it may justify having a dedicated seeder so you don't have to fiddle with calibration when you are trying to plant in a busy spring before rains!*

9 OTHER SEASONAL MANAGEMENT

Of course, gardens require many other tasks for seasonal management: routine cultivation, irrigation, harvest, etc. Permabeds should be either in cover crop, mulched, or under weed barrier for the winter months, rainy season, and the middle of summer. Row-based planting systems make all seasonal work efficient and effective, especially for tasks that are mechanized and use a two-wheel tractor!

DESIGN BOX

123 PLANTING METHOD

Permabeds should be planted using a determined number of variations of equidistant rows. This improves all mechanized tasks: seeding, cultivation, and more, because straight, parallel, and equidistant rows can be managed with equipment that is set up with similar spacing. The **123 Planting Method** primarily uses five variations of row quantity centering on *a constant middle row* (row #1) because all these rows are equidistant and all variations of row quantities (1 row for squash, 3 rows for kale, or 5 rows of radish) don't change. Instead, the spacing between the rows is adjusted by removing rows (5 rows of radish at 5" becomes three rows of kale at 10" by simply not planting the middle rows). By adjusting spacing simply by removing or adding rows then, this logic translates to your equipment; systems can be permanently set up for a modular approach to bed and crop management. Row marker units, seeders, or cultivators are just added or removed from their toolbar setup depending on if you are planting all 5 rows, or 3, or 2, or 1. For instance, a 5-row seeder could seed all 5 rows, or you can leave some hoppers empty to seed just 3 rows, or 2, or just 1 row. ***Note:*** Which crops use different spacing depends on numerous variables: variety size, chosen maturity for harvest, the method for cultivation or mulching, etc.

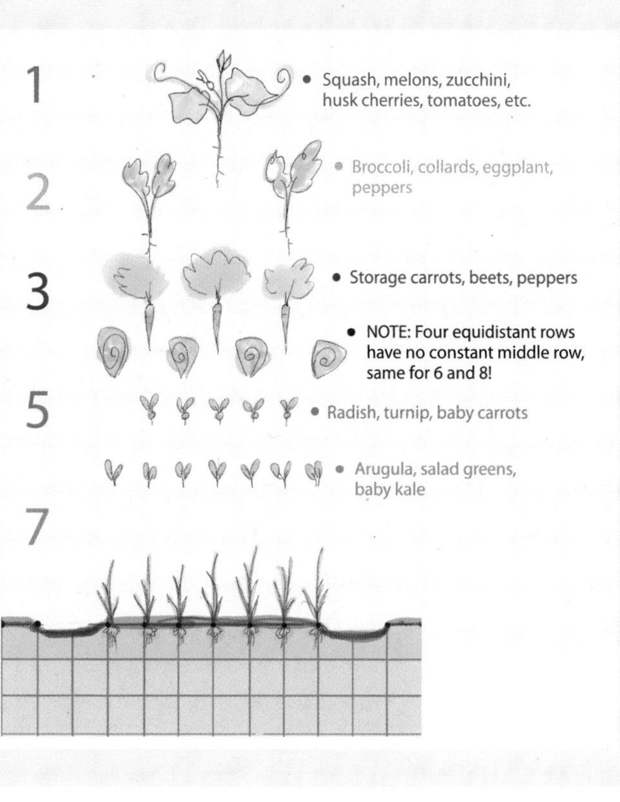

Chapter 8
Tractor Maintenance and Care

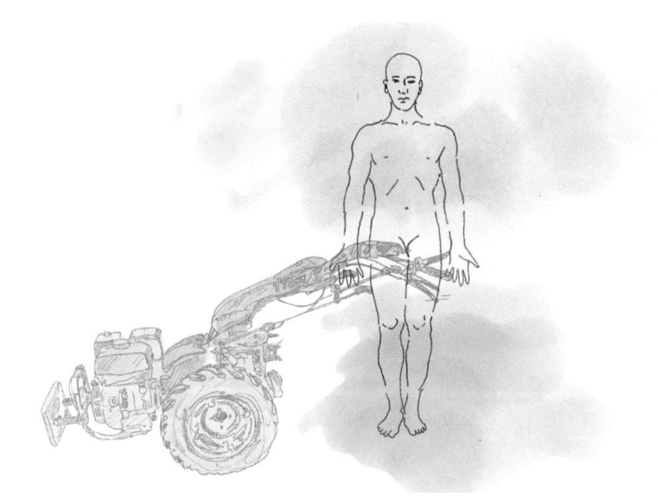

When it comes to maintenance and general operational characteristics, you can think of the *body of the tractor* as being composed of essential parts, fluids, and connections—just like our bodies are. And, just like our bodies, we need to keep tractors running in tip-top condition. A two-wheel tractor is a machine made of separate parts, but if you care for the *whole system,* it will work for a long time.

TRACTOR MAINTENANCE SIMPLIFIED

Most routine maintenance can be done by the grower because two-wheel tractor size and simplicity, relative to larger equipment, make them approachable for such jobs as engine oil changes and switching tires. Two-wheel tractors don't need a lot of work to keep them in good condition. Maintenance protocols (Figure 38) should be adopted from the get-go, and the needs of a particular tractor or implement should be considered for any equipment purchase and included in budgets at the start.

MAINTENANCE PROTOCOL PRINCIPLES

Maintenance can be understood holistically (Figure 38), but remember that the critical resource for all equipment and tractor maintenance is your *Owner's Manual.* This should be reviewed often for implement- and tractor-specific maintenance instructions. The principles below are to help create a mindset of maintenance rather than replace reading your manuals!

1. **Operation, Storage, Care:** Without proper operation, storage, and care, your tractor will eventually fail. Correct operation requires understanding the proper use of tractors and implements to avoid excessive wear and breaks. Proper storage is essential to make sure low seasons don't result in issues like condensation in your gas tank or cracked tires. Overall care and maintenance should be a matter of routine.
2. **Tools, Supplies, Space, and a Record System:** These are all needed to make sure you can fulfill protocol when needed. Proper tools can be arranged as a kit for easy access (with wrench sizes specific to your equipment, funnels suitable for different oil changes, etc.). Have all necessary supplies readily available for routine maintenance. Have a space set up with a workbench, plywood drop sheet for oil spills, etc. A record system helps keep you organized; this can consist of a shop booklet and/or spreadsheet software on a laptop.
3. **Checking, Knowing, and Timing:** You need to routinely *check* your tractor systems (like oil levels), *know* how to make correct adjustments (add to, or change it properly), and do it with the right *timing* (hours between oil changes). Hour meters can be purchased to help account for hours of operation, and notebook or computer records can help remind you of maintenance schedules.
4. All the tasks rely on the above supportive framework. *Examples:* **ADD** fuel stabilizer and **FILL** tank with gas before winter to prevent condensation. **CHARGE** battery and keep terminals **CLEAN**. **LUBRICATE** control cables and **ADJUST** when needed. **GREASE** the male tang of the quick connect, keep the PTO **CLEAN** and **LUBRICATE** the base of the vertical portion of the PTO engagement rod. **CHANGE** transmission fluid (and the filter for PowerSafe models—but *hand-tighten only*). **ADD** silicon sealant to transmission drain plug threads, and **TIGHTEN**.
5. Occasionally—and more often if the above instructions aren't followed—there will be repairs and part replacements.

FIGURE 37: MAINTENANCE PROTOCOL

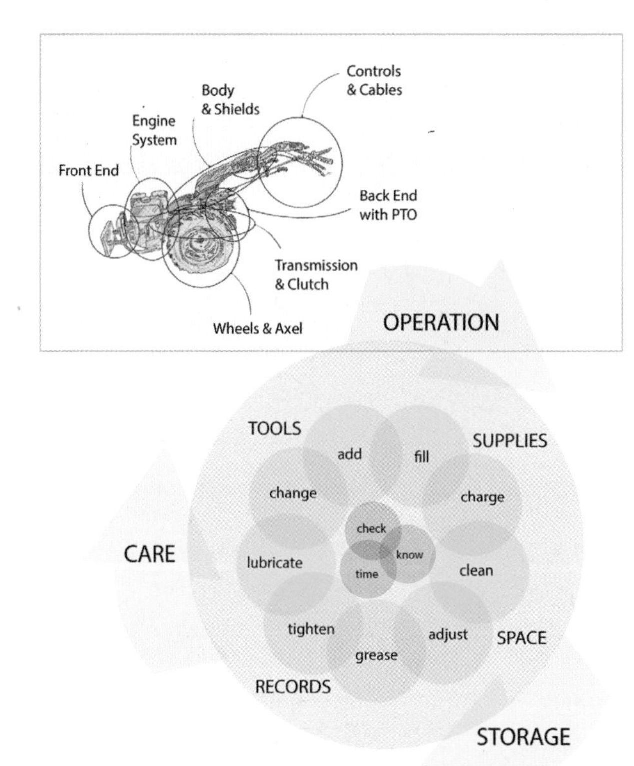

ESSENTIAL TOOLS FOR MAINTENANCE

1. **Tool kit bag and designated shelf:** Keep your gear in one place—ready for work in the field, if necessary.
2. **Full or basic set of metric wrenches and sockets** (8mm–24mm) for adjustments and adding extensions.
3. **Extra wrenches:** The most common sizes are 10mm, 13mm, 17mm, and 19mm, but you should find out exactly which wrenches are needed for regular adjustments and keep them handy with your tractor kit.
4. **Long, thin-tipped flat head screwdriver:** Can be used to free up a tight transmission dipstick cap by inserting and twisting between the bottom

ridge and transmission cover. ***Note:*** Generally useful when something is stuck.

5. **Grease gun:** For applying grease to any and all grease nipples.
6. **Small filter spanner for filter removal on Power-Safe models.** If you have several two-wheel tractors (which you very well may, if you have a larger operation), a small filter spanner can be handy for quick filter changes. In any case, the filter should only be hand-tight snug; you should be able to remove it by hand. ***Remember:*** When installing a new filter, apply a film of fluid to the rubber gasket and hand-tighten only.
7. **13⁄16" spark plug removal socket or tool** for spark plug removal. (There is a spark plug remover tool available, but a deep socket wrench will work, too.)
8. **White write-on all-surface markers:** Write the date of the last check or change directly on housing—OR keep records.
9. **Small hammer:** For gently tapping in quick-connect drop pins when they stick. ***Note:*** Pins shouldn't stick if the quick hitch is properly installed and the ground is level.
10. **Small jack:** Lifts up the tractor for wheel changes.
11. **Oil pans:** You should have 1 qt and 3 qt capacity pans for changing oils.
12. **Long neck oil funnel**: Useful for oil filling in hard-to-reach areas because of the long and adjustable neck.
13. **Wooden blocks:** (4 × 4 or 6 × 6) help with implement changes.
14. **Plywood drop sheet (not pictured):** Keeps the floor clean in case of spills and is easier on knees when working in a garage.

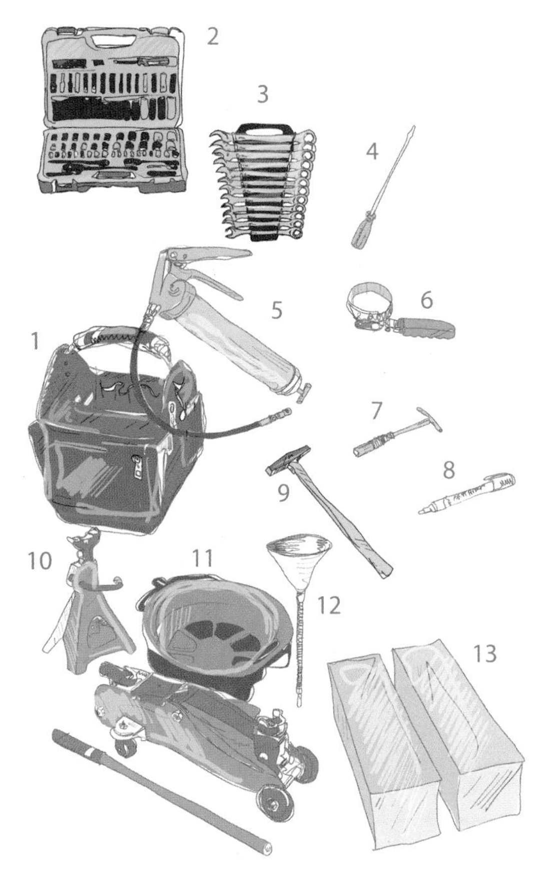

Jacking up the tractor to add extensions.

ESSENTIAL MAINTENANCE SUPPLIES

1. Scott Shop towels (general purpose): 1 roll of 55 sheets costs about $3.00. Keep them handy for cleaning tractor grease and wiping dipsticks clean for oil checking.
2. Permatex Fast Orange hand cleaner: One 7.5 oz container costs about $4.00. It's great for cleaning grease off hands.
3. Castrol Pyroplex Blue #2 (Water-Resistant EP Grease): A 14 oz tube costs about $7.00. Use it for greasing nipples on implements.
4. Emery cloth: One 8" × 10" sheet costs about $2.00. Useful for cleaning up any rust on PTO shafts, etc.
5. Scotch-Brite hand pads: Two 4" × 5" pads cost about $2: Use for cleaning rust and debris.
6. Clean-R-Carb Carburetor Cleaner: One 12 oz can costs about $14. Used to maintain the carburetor and act as a solvent for sediment cup rinsing.
7. Kleen Slip Silicone Lubricant/Deep Penetration Lubricant: One 12 oz can costs about $10. Use it to maintain all tractor cables (such as brake and clutch cables).
8. STA-BIL Fuel Stabilizer: One 8 oz container costs about $9. Use it to keep gas fresh over winter; a full tank helps reduce condensation and rust over winter.

Having the right products on hand is half the battle. Many of these are used in small amounts and will last years, and some only need to be purchased occasionally for oil changes.

9. Kleen-flo Honey Goo XX/All Purpose Penetration lubricant: A 350 g can costs about $7. Good general penetration lubricant for screws/bolts when anything is being removed or added.
10. Wurth HHS 2000 Adhesive Lubricant: One 330 g can costs about $22. Used to protect the PTO shaft and other exposed metal for long periods. Nice, because when it dries, it leaves a protective coating without being thick and messy
11. Kleen-flo Gasket Maker/Silicone Sealant: An 85 g tube costs about $8. Use it for resealing the transmission oil drain plug to make sure it stays on stay tight and clean. ***Note:*** This is for standard transmissions only, don't use on PowerSafe models.

MAINTENANCE TIPS

GENERAL:

- **Clean the tractor before maintenance:** Blow off any dust and clean tractor surfaces before maintenance.
- **Plan maintenance:** More involved maintenance tasks are best done at the beginning and end of the season.
- **Do routine checks** before each operation and new periods of heavy work.
- **Don't smoke** or have an open flame anywhere nearby when maintaining a tractor.
- **Wear proper safety gear** and maintain an ergonomic position. Protect your hands and knees.
- Have **clean-up supplies** on hand for spills.

ENGINE:

- **Check engine oil** every 4 hours of use.
- **Change engine oil** after an initial 20-hour "break-in period," then every 100 hours. The engine oil capacity is 1.2 US qt (1.1 liters) for most models.
- **Fuel line:** Honda recommends checking and replacing each season, if necessary.
- **Sediment cup:** Honda recommends cleaning every 6 months or every 100 hours.

- **Spark plugs:** Honda recommends cleaning and readjusting spark plugs every 6 months or every 100 hours. If the engine isn't functioning properly, spark plug problems are a likely cause. Consult a local Honda small-engine garage or dealer.
- **Fuel tank:** Honda recommends a professional cleaning every 300 hours.
- **Valve clearance:** Honda recommends a professional cleaning every 300 hours.
- **The air filter/cleaner** should be checked before each use. Honda recommends cleaning it every 3 months or every 50 hours, more often in dusty conditions.

TRANSMISSION:

- **Check transmission fluid** every 50 hours or before any period of new work (spring, summer, fall).
- **Change transmission or UTF fluid** as needed (say, every 100 hours).
- **Change UTF filter:** For hydrostatic models, it is recommended to change the oil filter after the first 30 hours and then with every oil change.

OTHER COMPONENTS:

- **Maintain cables** at end of the season by applying a silicone lubricant.
- **Maintain nuts and bolts:** Check for tightness. Apply lubricant to any older and rusty bolts when loosening or tightening them.
- **Sealant is used** after transmission fluid is changed. Use silicone sealant on drain plugs and put a little oil on the rubber oil ring of the dipstick at the fill caps after the job is finished (no dirt should ever get in here).
- **Keep the body guards (those plastic shields on tractor body) tightly fitted** to protect the tractor frame and cable guards from UV. These fiberglass shields can have lithium grease applied occasionally to the inside of the rubber bushings that push in and hold the shields in place. To best maintain the color of the shield itself, park the tractor indoors when not in use.
- **The male tang of the quick coupler must be protected** with a quality spray lubricant or grease (or even an anti-seize compound) between uses and while in storage. A rubber cover is also available to protect the tang in storage.

- **Tighten and adjust cables when needed, and check the tightness of all lever-securing hardware:** Levers and associated cables will occasionally need to be tightened. Levers to check: clutch, differential, reverse, emergency stop, handlebar rotation, and right and left braking. Keep cables gathered and not too freely hanging using small zip ties.
- **Electrical:** The electrical connections between the OPC lever and the engine (on non-PowerSafe models) require care to ensure wires are protected and aren't ripped loose. Small zip ties help if wires are hanging too freely.
- **Battery:** Battery terminals should be greased to prevent corrosion and keep the cables well-connected, with bolts tightened. *Never over-tighten.* Batteries are best removed and stored somewhere warm over winter if the tractor is not going to be used in the wintertime. (Of course, not all models have batteries.)

WINTER STORAGE:

- **Fill the gasoline tank** till full and add a fuel stabilizer to the gasoline tank, then run the engine for 10–15 minutes to circulate the now-stabilized gas throughout the system.
- **Top up the tires** to full pressure and park the tractor on a piece of plywood so the tires aren't in constant contact with a cold concrete slab or wet gravel floor.
- **Apply grease** to the battery terminals. Batteries can be disconnected and brought indoors to a basement or kept in a heated garage. They should be trickle-charged if you are going to leave them for a long time. When batteries have low voltage and the temperatures are freezing, batteries can be damaged.

We want two-wheel tractors to keep well over the winter and be ready to go come spring! Do all the tasks shown in these pictures, making sure to grease all necessary components. Winter preparation, like adding a fuel additive, is an *essential* part of maintenance. Having your parking brake on or putting the tractor in gear while doing maintenance keeps the tractor firmly rooted while you push/pull when turning nuts and bolts, and making other adjustments.

FIGURE 38: OTHER MAINTENANCE

Spark plug

Lubricating a rusted male quick coupler tang

Cleaning a rusty male tang

Lubricating cables

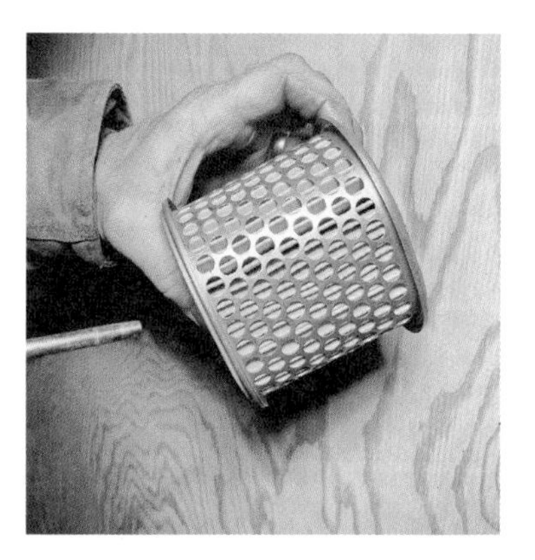

Cleaning the air filter

Checking air filter's outer oil cover

Adding fuel stabalizer

Greasing the flail mower

Parking brake is ON for maintenance

CHECKING AND CHANGING HONDA ENGINE OIL

We won't go step-by-step through all tractor maintenance for all models. But here is the process for checking and changing the oil on a Honda engine (see Figure 39). Many two-wheel tractors come with Honda engines, including the GX240, GX270, GX340, and GX390 (popular on two-wheel tractors). Check your engine manual for more details. ***Note:*** For all your maintenance needs, there is a logical procedure you can learn and follow.

- You should check the oil after every 4 hours of operation.
  - Remove the dipstick, wipe it clean, and insert it into the oil filler hole.
  - Remove the dipstick and check the oil level.
  - If the oil level is low, fill oil up to the top of the oil filler neck.
- Change the oil after a 20-hour "break-in" period, then after every 100 hours of use.
  - Run the engine for 5–15 minutes to warm the oil before draining. The last dregs of oil are the dirtiest, and they won't come out unless the oil is warmed.
  - The oil discharge plug is located at the bottom of the engine block and can be cleaned with a towel and then removed with a ½" socket wrench.
  - Allow the oil to drain into a drain pan and dispose of the oil according to local regulations.
  - Use a block of wood under the opposite wheel to help those last dirty dregs of oil to come out.
  - Return the tractor to a level position.
  - Clean and replace the drain plug.
  - Clean filling cap and then remove.
  - Using a flexible funnel (others won't reach) inserted into the filler neck, add the specified amount of engine oil (or until it is completely full).
  - Honda recommends a Honda 4-stroke oil or equivalent high-quality motor oil (use a high-detergent type to prevent rust, corrosion, and oxidation).
  - 10W-30 is recommended for most users as an all-temperature engine oil.

DESIGN BOX
UNDERSTANDING DIFFERENT OILS

What's in an engine oil's name? The Society of Automotive Engineers (SAE) developed the standards for engine oil. Let's consider *SAE 10W-30.* The W stands for "Winter." The first number is the winter viscosity and the second is summer viscosity. A less viscous oil moves around the engine crankcase more easily, reaching and lubricating everything evenly. In colder temperatures, a lower viscosity helps because fluids are already more sluggish in colder temperatures. Think: low temperature requires low viscosity. So, because low temperatures require low winter viscosity (W), 0W-20 is better than a 10W-30 for winter.

FIGURE 39: CHANGING THE ENGINE OIL ON A POWERSAFE BCS TRACTOR

Checking engine oil with dipstick

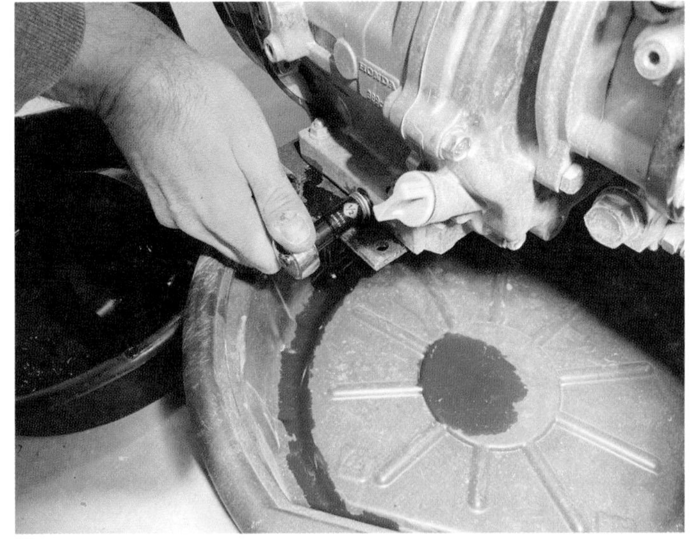

Removing engine oil drain plug

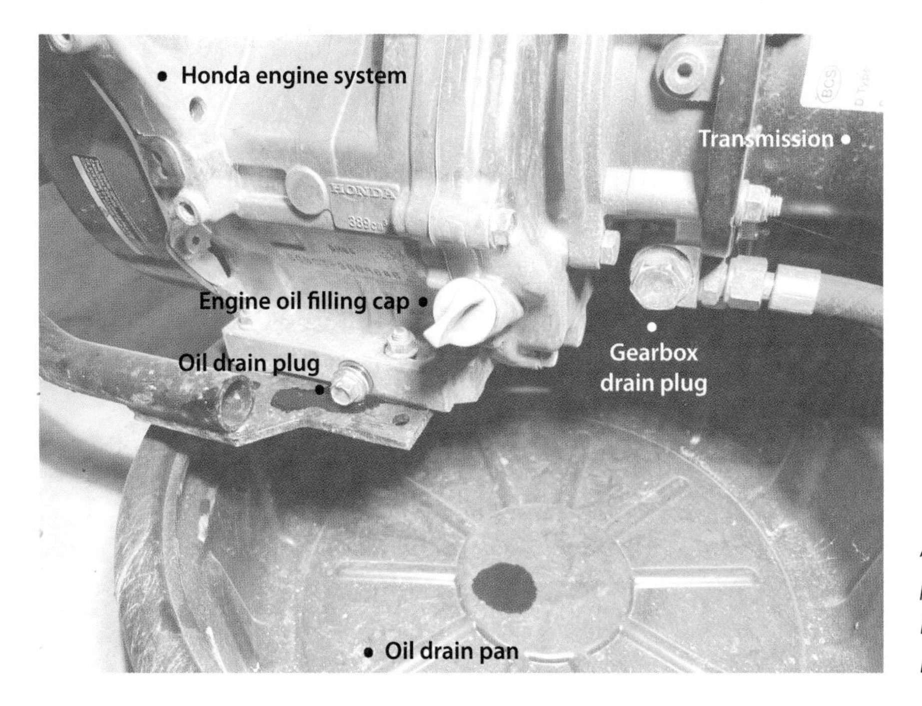

A look at the engine oil drain plugs and fill caps. ***Note****: These images show a PowerSafe model.*

TWO-WHEELING INTO THE FUTURE

And so, our journey has come to an end, to begin again.

CHANGES TO COME

In the future, two-wheel tractors will undoubtedly change. Already, a shift to hydrostatic and electric is in the works. But for the foreseeable future, there will still be a demand for all-gear transmissions and the familiar and powerful gas engine. It is the diversity of options and solutions that will lead us forward, and the two-wheel tractor—alongside humble hand tools—will continue to be the means of choice for progressive, small-scale land management!

The future could also bring increased access to equipment rental for two-wheel tractors from tool libraries, farm equipment cooperatives, and specialized equipment retailers dedicated to small-scale growers—so growers can purchase some equipment guilds that suit primary uses, while also being able to access a variety of other equipment for short-term jobs. This will strengthen the multi-functional nature of this tractor and further improve the affordability of its initial purchase. The role of small equipment innovation and manufacturing will also increase as need rises. Leaders in agriculture and landscaping will be right there helping innovate the newest solutions to forward-thinking garden, farm, and landscape equipment design and techniques.

NEW USERS

Many key industries and enterprises surprisingly *haven't* adopted two-wheel tractors and still rely on single-purpose machines (landscaper equipment tends to be single-purpose), or they are still using four-wheel tractors even when small-scale equipment would be more suitable (true of many Homesteaders). Despite the obvious advantages of two-wheel tractor equipment, there is a pervasive mismatching of project and equipment-scales. Many farmers still rely on four-wheel tractors for jobs better suited to two-wheels—like pulling a three-row seeder or managing in their greenhouses.

Why? Who killed the two-wheel tractor? Remember, in the early and mid-20th century, two-wheel tractors were ubiquitous for DIY and professional growers. Yet, maybe a better question to ask is: **Who killed small-scale food production?** Or even better still: **Who will help to bring it back?**

GROWING DIVERSITY ON TWO WHEELS

Today, the concern around food quality, quantity, and cost is evermore in our minds, and small-scale solutions for food production will increasingly be relevant. The global upheaval in the last three years is introducing a new generation (and reminding an older one) of the uncertainty of markets, flaky trade routes, and the need for resilience. We are questioning the attainability of the quality of life we all deserve. Small-scale land management of gardens, edible landscapes, and crops need earthworking, mowing, chipping, and chopping. Two-wheel tractors can help meet these needs. The affordability, maneuverability, and versatility of two-wheel tractors makes them equally available to professionals and DIY property owners and across urban, suburban, and rural settings. As such, this equipment is ideally suited to help catalyze the transition of our food system to local sustainable agriculture and resilient homesteading, as well as transform public lands to edible diversity conservation in parks, community gardens, and yards.

Remember, with only a single engine to maintain, these tractors can perform a diversity of tasks: from plowing, bed forming, and compost spreading to seeding, cultivation, and mowing, and from brush mowing to wood chopping and chipping. Equipment and techniques strategically linked can meet projects *where they are at and grow with them as they scale up*—even *bridging* the important gap between a start-up enterprise and going pro. Equipment guilds help budgeting by organizing around scale phases, enterprises, and intended static-scale. This extends purchases over time and avoids missteps in equipment acquisition. With sound decision-making at home, on the farm, and across entire landscapes, stewards can budget for successful enterprises.

The leading edge of our food system is about streamlining the transition of land to edible biodiversity and turning DIY enterprises into profitable small-scale productions. Proper equipment design gives growers the power to manage highly productive systems—such as the Permabed System for edible biodiversity. Here, equipment tasks are scheduled alongside guild crop rotations to maximize beneficial activities like annual/perennial patterning, in situ mulching, and the Compost-a-Path Method, boosting short-term yield and long-term soil health. Diversified growing systems need scale-suitable equipment, and edible biodiversity will be the guiding light as our society moves forward into the middle of this century in search of profit resilience.

The two-wheel tractor is here to stay, and the future holds excitement for those who like getting in the dirt, woods, and fields.

So, let's *get two-wheeling into the future, with our feet firmly on the ground.*

Grow on!
Zach Loeks

GLOSSARY OF COMMON TERMS AND JARGON

Here are some important terms and themes that anchor the content of this book.

Two-Wheel Tractor: Tractor with a single axel and two wheels that can be connected to many different implements. This is also referred to as a *2w-tractor.*

M2w-Tractor: In this book, M2w-tractor refers to multi-functional two-wheel tractors, such as the BCS with PTO, which has a larger engine and fully adjustable handlebars. Row crop tractors are referred to by their name, while all two-wheel tractors are sometimes collectively referred to as 2w-tractors. This simply saves space.

Compost-a-Path: A method for managing crop debris and cover crops as in situ fertility and mulch.

Crop Successions: A series of routine seasonal seeding/transplanting events for continuous supply. For example, seeding spinach every week in spring and carrots every two weeks through mid-summer.

Equipment Guild: A complement of equipment that helps complete a series of garden tasks as a whole operation cycle (such as reforming and preparing new beds for seeding crops). Using equipment guilds is an important strategy for equipment selection and scaling-up enterprises. (See Chapters 4 and 5.)

Equipment: Equipment can include hand tools and tractor implements (tractor attachments).

Guild Enterprise Production: This is when a farm or homestead has a mix of different enterprises that work together. Overall, *guild enterprise production* means the projects are designed to share equipment, balance labor, inform decision-making, and cycle waste as inputs.

Guild: This term is used broadly to refer to any assembly of mutually beneficial entities. This includes plants, both annuals and perennials (the common concept for growers), but also equipment and enterprises. Such as a crop guild, or guild crop rotation.

Holistic Scale Principles: These include key aspects of scale that help with decision-making (see Figure 1).

Multi-functional: This refers to an item or action that has many uses. The more functions a tractor or specific implement can perform, the better its cost recovery.

Permabed System: This is a land management design for intensive and extensive growers of annuals and/or perennials—and even animals—that uses *ecosystem design*. This system employs a complement of landscape layout, guild design, crop rotation, and input/output management techniques blended as a complete multi-year management cycle with a set of Permabed system principles.

Permabeds: Permanent raised garden beds that are never destroyed and only reformed. Used for intensive growing with regenerative soil practices and as a means of organizing and growing a diversity of crops.

Profit Resilience: When an enterprise is both profitable *and* resilient in terms of socio-economic and environmental gains in both the short- and long-term.

Scale Phases: All land-based enterprises undergo a series of scale phases: *start-up*, *scale-up*, and *pro-up* are scale phases that are outlined in this book as a way of approaching logical equipment-use decision-making. They mark a three-phase approach to reach a planned and desired scale and steady-state operation cycles.

Scale: *Scale* is an elusive term. In this book it is understood as a relationship between multiple interrelated aspects of land management, including, but not limited to: acreage, profit, labor, equipment, management style, etc. Equipment selection must be based on proper scale analysis and planning.

Scale-Suitability: Defines plans, design, and equipment that are practically chosen to fit your scale of operations.

Small-Scale Farming: Management of actual production acreage from between ¼ to 15 acres and primarily using intensive techniques and smaller equipment types. Usually done for high profit per acre, but the concept of "profit" must be understood holistically.

Static (Production) Scale: This is the final scale that is reached by growers who set their intention on a certain acreage, management style, income, and equipment complement as they pro-up.

INDEX

Note: *italic* page numbers refer to figures.